物理化学实验

（双语）

主　编　余建国
副主编　曲玉宁　安会琴　王丽丽
　　　　贺晓凌　严　峰　王　兵

哈尔滨工程大学出版社
Harbin Engineering University Press

内 容 简 介

本书根据工科类院校本科基础化学教学的要求编写,着眼于培养有综合素质及创新意识的工科人才。全书力求全面地反映出工科类院校物理化学实验教材的结构与内容,涉及范围广,应用性强。书中选编了基础实验、综合性实验和设计型实验三类共22个实验项目,内容涉及热力学、相平衡、化学平衡、电化学、化学反应动力学、表面现象及胶体化学等。每个实验后编写了与该实验相关的背景阅读及实验技术等相关内容,以提高学生兴趣及扩展其视野。仪器部分详细地介绍了物理化学实验涉及的一些仪器的基本原理及使用方法。附录部分收录了大量必要的物理化学实验数据。本书在内容安排上适应当前教学改革的需要,既有传统的验证实验,也有反映现代物理化学的新进展、新技术及与应用密切结合的实验,具有基础性、应用性和综合性等特点,同时在基础化学实验与专业实验衔接方面也做了一些尝试。

本书可作为高等院校化学、化工、环境、制药及材料等专业的教材,也可作为相关专业技术人员的参考书。

图书在版编目(CIP)数据

物理化学实验:双语 / 余建国主编. — 哈尔滨:
哈尔滨工程大学出版社, 2021.7(2025.1 重印)
ISBN 978 – 7 – 5661 – 3123 – 2

Ⅰ. ①物… Ⅱ. ①余… Ⅲ. ①物理化学 – 化学实验 –
双语教学 – 高等学校 – 教材 Ⅳ. ①O64 – 33

中国版本图书馆 CIP 数据核字(2021)第 137632 号

物理化学实验(双语)
WULI HUAXUE SHIYAN(SHUANGYU)

选题策划 刘凯元
责任编辑 刘凯元
封面设计 李海波

出版发行	哈尔滨工程大学出版社
社　　址	哈尔滨市南岗区南通大街 145 号
邮政编码	150001
发行电话	0451 – 82519328
传　　真	0451 – 82519699
经　　销	新华书店
印　　刷	哈尔滨午阳印刷有限公司
开　　本	787 mm×1 092 mm　1/16
印　　张	15.25
字　　数	383 千字
版　　次	2021 年 7 月第 1 版
印　　次	2025 年 1 月第 2 次印刷
定　　价	40.00 元

http://www.hrbeupress.com
E-mail:heupress@ hrbeu.edu.cn

前 言

物理化学实验是高等院校化学、化工、应用化学、轻化工程、材料、环境、石油、制药等工科类专业的一门重要的实验课。随着物理化学研究方法的形成和发展，以掌握物理化学现代实验技术和方法为主的教学目的也愈来愈向培养学生的综合能力和科学研究能力发展。为进一步贯彻教育部全面提高教学质量、培养高素质人才及加强教材建设的精神，本书在传承以往工作的基础上，参考了国内外大量的相关资料，综合了化学领域中各学科所需的基本研究手段和方法，既包括物理化学基础实验部分，又涵盖了综合性实验和设计型实验，还设计了部分英文实验内容，力图与国际接轨，使学生通过训练培养创新思维能力与初步进行科学研究的能力。

本书注重实验方法与手段的更新与发展，力求反映物理化学新进展、新技术，并与应用紧密结合。全书共5章，实验部分按照基础实验、综合性实验和设计型实验的不同层次展开，实验内容涉及热力学、相平衡、化学平衡、电化学、化学反应动力学、表面现象及胶体化学等。每个实验后编写了与该实验相关的背景阅读及实验技术等相关内容，以提高学生兴趣及扩展其视野。仪器部分详细地介绍了物理化学实验涉及的一些仪器的基本原理及使用方法。附录部分收录了大量必要的物理化学实验数据。

本教材在实验内容与学时安排上具有可操作性，适用于工科类高等院校与化学有关专业的不同层次的物理化学实验教学使用，也可供科研人员参考。书中英文实验内容，可供外国留学人员参考阅读。特别地，本书前8个基础实验附有部分操作及实验原理的短视频（观看该视频需泛亚平台账号），其链接网址：http://mooc1.chaoxing.com/course/99697700.html，可登录进行观看。

参加本书编写工作的有余建国、曲玉宁、安会琴、王丽丽、贺晓凌、严峰、王兵，全书由余建国统稿。本书的编写还得益于天津工业大学国际交流学院的立项资助以及兄弟院校的支持，在此深表谢意。由于编者水平有限，疏漏之处在所难免，希望读者不吝指正。

编 者
2020年12月

目　　录

第1章　绪论 ··· 1

　1.1　物理化学实验的目的和要求 ·· 1

　1.2　物理化学实验中的误差与数据表达 ·································· 3

　1.3　用计算机处理实验数据和表达实验结果 ···························· 11

第2章　基础实验 ·· 27

　实验1　恒温水浴调节和黏度的测定 ······································ 27

　Experiment 1　Determination of Liquid Viscosity ······················ 33

　实验2　二元液系的气液平衡相图 ··· 35

　Experiment 2　Phase Diagram of Liquid-Vapor Equilibrium in a Binary System ·········· 42

　实验3　液体饱和蒸气压的测定 ·· 46

　Experiment 3　Determination of Saturated Vapor Pressure of a Pure Liquid ··········· 53

　实验4　氨基甲酸铵分解反应平衡常数的测定 ··························· 57

　Experiment 4　Determination of Equilibrium Constants for Ammonium Carbamate

　　　　　　　　Decomposition Reaction ································· 64

　实验5　乙酸乙酯皂化反应速率常数的测定 ······························ 68

　Experiment 5　Determination of Rate Constant for the Saponification of Ethyl Acetate ······ 74

　实验6　蔗糖水解反应速率常数的测定 ···································· 77

　Experiment 6　Determination of Rate Constant for Hydrolysis of Sucrose ············ 85

　实验7　原电池电动势的测定 ·· 88

　Experiment 7　Determination of the Electromotive Force of Galvanic Cells ············ 98

　实验8　表面活性剂临界胶团浓度的测定 ································· 103

　Experiment 8　Determination for Critical Micelle Concentration of Surfactant ········· 112

　Experiment 9　Determination of the Rate Law for the Iodination of Acetone ··········· 116

　Experiment 10　Determination of Molar Mass by Freezing Point Depression ·········· 123

第3章　综合性实验 ··· 131

　实验11　B-Z振荡反应 ··· 131

　Experiment 11　B-Z Oscillation Reaction ································ 141

　实验12　铝阳极氧化膜电解着色 ·· 146

　Experiment 12　Electrolytic Coloring of Aluminum Anodic Oxide Film ············· 150

　实验13　溶胶和乳状液的制备及其性质 ··································· 154

实验 14　黏度法测定高聚物的摩尔质量 ·· 160

实验 15　溶液吸附法测定固体比表面积 ·· 166

Experiment 15　Determination of Solid Specific Surface Area by Solution Adsorption Method ·· 170

实验 16　蛋白质等电点的测定 ·· 174

Experiment 16　Determination of Protein Isoelectric Point ·· 179

第 4 章　设计型实验 ·· 184

实验 17　电还原草酸制备乙醛酸的方法 ·· 184

实验 18　电镀铜 ·· 188

实验 19　过氧化氢分解催化剂的制备及其性能比较 ·· 192

实验 20　植物色素热降解动力学参数的测定 ·· 195

Experiment 20　Determination of Kinetic Parameters of Thermal Degradation of Plant Pigments ·· 200

Experiment 21　Nano-TiO_2 Prepared by Sol-gel Method and Its Photocatalytic Activity ·· 204

Experiment 22　Determination of Pd/C Catalytic Performance for Hydrogen Evolution Reaction (HER) ·· 206

第 5 章　仪器 ·· 208

仪器 1　恒温水浴 ·· 208

仪器 2　阿贝折射仪 ·· 211

仪器 3　数字式差压计 ·· 214

仪器 4　电导率仪 ·· 216

仪器 5　旋光仪 ·· 220

仪器 6　EM-2A 型数字式电子电位差计 ·· 224

仪器 7　JK99 全自动表面张力仪 ·· 225

仪器 8　JS94H 微电泳仪 ·· 227

附录　常用数据表 ·· 230

参考文献 ·· 237

第1章 绪　　论

1.1　物理化学实验的目的和要求

物理化学实验是物理化学教学中的重要环节,目的是通过实验的手段,研究物质的物理化学性质,以及这些性质与化学反应之间的关系,从中形成规律性的认识,使学生掌握物理化学的有关理论、研究方法和实验技术,包括实验现象的记录、实验条件的选择、重要物理化学性能的测量、实验结果的分析和归纳等,从而增强解决实际化学问题的能力,加深对物理化学课程中某些重要的基本理论和基本概念的理解。

1.1.1　实验前的预习

在进行实验之前,必须做好充分准备,明确实验中每一步如何进行及为什么要这样做。这样才能较好地完成实验课的任务,杜绝原理上、方法上的错误,因为这些错误有时可能导致整个实验的失败。另外根据物理化学实验的特点,往往采取循环安排,有些实验在课堂讲授有关内容之前就要进行。因此,实验前充分进行预习,对于做好物理化学实验尤为重要。

预习时一般应做到仔细阅读实验教材,必要时参考教科书中的有关内容,学习实验方法、原理及如何使用仪器。要求明确实验目的,掌握实验所依据的基本原理,明确需要进行哪些测量,记录哪些数据,了解仪器的构造及操作并写出预习报告。预习报告用A4纸书写,报告中应包括实验名称、实验目的、简明原理、需要用到的实验仪器和实验条件、操作要点、注意事项,列出原始数据表,在实验前交指导教师检查。

1.1.2　实验过程

进入实验室必须遵守实验室各项规章制度。经教师检查提问,认为达到预习要求后才能进行实验操作。在整个实验过程中,应严格按实验操作规程仔细地进行操作。实验中仔细观察,客观记录数据,不能用铅笔或红笔记录,不能记在纸片上,原始数据不能涂覆,按规定修改记错、写错的数据。若有可能,可在实验过程中对实验结果进行初步计算或画出草图,以了解实验的进展。

实验结束前,应核对数据,并对最后结果进行估算,如有必要,可补测数据。实验结束,经教师检查后,拆卸实验装置。将原始数据登记在实验室指定的记录本上,记录实验时的室温、气压、天气等数据及实验日期、时间、实验合作者、指导教师等资料,整理、清洁仪器及实验台,然后离开实验室。

实验室内应保持安静,不得高声喧哗及任意走动,应严格遵守实验室安全守则,以保证实验顺利进行。

实验中应注意爱护仪器,节约药品。实验结束后,清洗并整理好仪器,在仪器使用登记本上写明仪器使用情况并签名,经教师检查后方可离开实验室。

1.1.3 实验报告

实验完毕,学生必须将原始记录交教师签名,然后正确处理数据,写出实验报告。写实验报告的目的有两个:一是向教师报告实验结果和对结果的分析;二是锻炼总结和表达实验结果的能力。要求每个参加实验的人都要写报告,以便及时总结和互相交流。

物理化学实验报告一般应包括实验目的、简明的实验原理、实验仪器和实验条件、具体操作方法、实验数据、结果处理、问题及讨论等。

实验目的应该用简单、明了的文字说明所用实验方法及研究对象。

实验仪器用简图表示,注明各部分的名称,若仪器很简单,这一项可以略去。

实验数据尽可能以表格形式表示,每一项标题应简单、准确,不要遗忘某些实验条件的记录,如室温、大气压力等。

在结果处理中应写出计算公式,注明公式中所需的已知常数的数值,注意各数值所用的单位。若计算结果较多时,最好也用表格形式表示,有时也可以将实验数据和结果处理合并为一项。

统一采用计算机软件作图,并用 A4 纸打印,附于实验报告相应位置。

讨论的内容应包括实验中观察到的特殊现象,以及关于原理、操作、仪器设计和实验误差等问题的分析。

写实验报告可以有自己的风格,但必须清楚而简要。简要并不是排除必要的细节,而是用最简练的语言完整地表达所要说明的问题。对于一些技术名词,使用必须规范。

书写实验报告时,要求开动脑筋、钻研问题、耐心计算、仔细编写,切忌粗枝大叶、字迹潦草。通过编写实验报告,达到加深理解实验内容、提高写作能力和培养严谨科学态度的目的。

1.1.4 设计型实验要求

设计型实验不同于综合性实验,它要求利用物理化学原理、实验方法和实验技术开展初步的科学研究,进行具有创新性的工作。所谓创新,不是重复别人已经做过的工作,而是针对自己感兴趣的课题,有所发现,有所发明,有所创造。研究结果可以作为学术论文在公开刊物上发表。

物理化学的理论和方法是化学各个学科的基础。对于化学来说,研究各种化学物质和材料的制备、性能以及它们与物质微观结构的关系是一条主线。平衡和速率是制备和性能研究中最基本的问题。物理化学的研究对象包括有机化学、无机化学、分析化学的各个分支,其研究方法如热力学方法、动力学方法、统计热力学方法及量子力学方法对化学各个学科都有重要应用。因此,研究创新型实验可以涉及不同的专业、不同的领域,其核心在于这类实验要体现物理化学的原理与方法。

这类实验的目的在于为学生提供一个开展科学研究的空间。通过这类实验,在老师的指导下,初步完成由学习到应用、由读书至进行科学研究的转变,全面提高学生的综合素质。这类实验要求学生在完成基础实验、综合性实验的基础上进行,时间可以较长,可以在课余时间进行,最终要求研究结果以学术论文的形式发表。这类实验可以按本部分提供的实验示例开展,也可由学生自拟题目,经指导教师同意后即可进行前期的准备工作。

任何一项科学研究都离不开现有的实验手段和方法。本部分给出一些关于电化学、催化化学及动力学研究的基本手段、方法及实验示例,供同学们参考和学习。希望在此基础上,在教师指导下开展一些创新型实验。

1.2 物理化学实验中的误差与数据表达

在物理化学实验过程中,通常是对某一系统的物理化学性质与系统发生化学反应之间的关系进行研究,以测量系统的某些物理量为基本内容,通过对所测得的实验数据进行分析和处理,从中获得重要规律。实验方法的可靠程度、所用仪器的精密度和实验者感官的限度等各方面条件的限制使得一切测量均带有误差,也就是测量值与真值之差。因此,要对误差产生的原因及其规律进行研究后,方可在合理的人力、物力支出条件下,获得可靠的实验结果。再通过实验数据的列表、作图、建立数学关系式等处理步骤,使实验结果变为有参考价值的资料,这在物理化学实验乃至科学研究中是十分重要的。

1.2.1 物理量的测量与单位

1. 物理量的测量

物理量的测量可分为直接测量与间接测量两种方式。将被测量的量直接与同一类量进行比较的方法称为直接测量。如用米尺量长度、秒表计时间、温度计测温度、天平称物质质量等。测量结果要由若干个直接测量的数据应用一些公式计算才能得到的方法称为间接测量,如黏度法测乙醇的相对黏度,就是用毛细管黏度计测出纯水和乙醇的流出时间,然后利用公式求得乙醇的相对黏度。物理化学实验的大多数测量问题是通过间接测量解决的。

2. 物理量的单位与数值的规定

物理化学实验数据的记录与表达一般采用国际单位制(SI)。我国的法定计量单位等效采用国际标准。有关 SI 和我国的国家标准的叙述及讨论可参阅有关文献、标准。

使用 SI 时应注意以下几点关于单位与数值的规定:

(1) 组合单位相乘时应该用圆点或空格,不用乘号。如密度单位可写成 kg/m^3 或 $kg \cdot m^{-3}$,不可写成 $kg \times m^{-3}$。

(2) 组合单位中不能用一条以上的斜线,不可写成 J/K/mol 等错误形式。

(3) 对于分子无量纲、分母有量纲的组合单位,一般用负幂形式表示。如 K^{-1}、s^{-1} 不可写成 1/K、1/s。

(4) 任何物理量的单位符号应放在整个数值的后面,如 1.52 m 不可写作 1m52。

(5) 不得使用重叠的词头,如 nm(纳米)、Mg(兆克)不可写作 mμm(毫微米)、kkg(千千克)。

(6) 数值相乘时,为避免与小数点相混,应采用乘号而不用圆点。如 2.58×6.17 不可写作 2.58·6.17。

(7) 组合单位中,中文名称的写法与读法应与单位一致,如比热单位是 $J/(kg \cdot K)$,即 "焦耳每千克开尔文",不应写或读为 "每千克开尔文焦耳"。

1.2.2 误差的分类

误差按其性质可分为系统误差、偶然误差和过失误差三种。

1. 系统误差

在相同条件下,多次测量同一量时,误差的绝对值和符号保持恒定,或在条件改变时,按某一确定规律变化的误差称为系统误差。系统误差产生的原因有如下几点:

(1) 仪器、药品不良,如电表零点偏差、温度计刻度不准、药品纯度不够等;

(2) 实验方法方面的缺陷,如使用了近似公式;

(3) 环境方面的影响,使测量数据不是偏大就是偏小,如折射率、旋光度、光密度等均与温度有关;

(4) 操作者的习惯,如观察视线偏高或偏低,对颜色的敏感程度差异等。

系统误差可分为不变系统误差和可变系统误差。在整个测量过程中,符号和大小固定不变的误差称为不变系统误差。例如,天平砝码未经校正;某 100 mL 容量瓶的实际容积为 101 mL;在使用中由于未加校正而引入固定的 +1 mL 系统误差。可变系统误差是随测量值或时间的变化,误差值和符号也按一定规律变化的误差。改变实验条件可以发现系统误差的存在,针对产生的原因可采取措施将其消除。系统误差产生的原因不能完全知道,通常可采用几种不同的实验技术,或采用不同的实验方法,或改变实验条件、调整仪器、提高试剂的纯度等,以便确定有系统误差存在,并确定其性质,然后设法消除或减少。

2. 偶然误差(随机误差)

在相同条件下多次测量同一量时,误差的绝对值时大时小,符号时正时负,但随着测量次数的增加,其平均值趋近于零,即具有抵偿性,此类误差称为偶然误差。它产生的原因并不确定,一般是由环境条件的改变(如大气压、温度的波动),操作者感官分辨能力的限制(如对仪器最小分度以内的读数难以读准确等)所致。它在实验中总是存在的,无法完全避免,但它服从概率分布。如在同一条件下对同一物理量多次测量时,会发现数据符合如图 1-2-1 所示的正态分布规律,该曲线称为偶然误差的正态分布曲线。其横轴表示偶然误差 σ 为无限多次测量所得的标准误差,纵坐标表示偶然误差出现的次数 N。

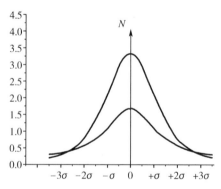

图 1-2-1 偶然误差的正态分布曲线

(1) 偶然误差的规律

由图 1-2-1 可见偶然误差的规律如下。

①对称性:绝对值相等的正误差、负误差出现的概率几乎相等,正态分布曲线以 y 轴对称。

②单峰性:小误差出现的概率大,大误差出现的概率小,很大误差出现的概率接近于零。

③有界性:在一定测量条件下的有限次测量值中,偏差的绝对值不会超过某一界限。

(2)偶然误差的表达方法

①平均误差

$$\delta = \frac{\sum |d_i|}{n}$$

式中,d_i 为测量值 x_i 与算术平均值 $\bar{x}$ 之差,$\bar{x} = \frac{\sum |x_i|}{n}$($i = 1,2,\cdots,n$,以下同);$n$ 为测量次数。

②标准误差(均方根偏差)

$$\sigma = \sqrt{\frac{\sum d_i^2}{n-1}}$$

式中,$\sum d_i^2 = (x_1 - \bar{x})^2 + (x_2 - \bar{x})^2 + \cdots + (x_n - \bar{x})^2$。

③或然误差

$$p = 0.675\sigma$$

或然误差 p 的意义是在一组测量中若不计正负号,误差大于 p 的测量值与误差小于 p 的测量值将各占测量次数的 50%,即误差落在 $+p$ 与 $-p$ 之间的测量次数占总测量数的一半。

平均误差的优点是计算方便,仅用这种误差表示时,可能会把质量不高的测量掩盖住。标准误差对一组测量中的较大误差或较小误差感觉比较灵敏,因此它是表示精度的较好方法。在近代科学中一般采用标准误差。此外,为了表达测量的精度,又有绝对误差、相对误差两种表达方法。

(3)测量结果表示方法

绝对误差表示测量值与真值的接近程度,即测量的准确度。其表示法为 $\bar{x} \pm \delta$ 或 $\bar{x} \pm \sigma$,其中 δ 和 σ 分别为平均偏差和标准偏差,一般以一位数字(最多两位)表示。

相对误差表示测量值的精密度,即各次测量值相互靠近的程度。其表示法为平均相对误差 $= \pm \frac{\delta}{\bar{x}} \times 100\%$,标准相对误差 $= \pm \frac{\sigma}{\bar{x}} \times 100\%$。测量结果表示为 $\bar{x} \pm$ 平均相对误差或 $\bar{x} \pm$ 标准相对误差。

3. 过失误差(粗差)

过失误差是一种明显歪曲实验结果的误差。它无规律可循,不属于测量误差范畴,是由操作者读错实验数据、记录错误、计算错误引起的误差。只要实验者加强责任心,此类误差可以避免。发现有此种误差产生,所得数据应予以剔除。

1.2.3 精密度与准确度

精密度是指测量值重复性的大小,偶然误差小,数据重复性就好,测量的精密度就高。准确度是指测量值与真值的符合程度,系统误差与偶然误差都小,测量值的准确度就高。一个精度好的测量,其准确度不一定很好,但要得到高的准确度就必须有高精度的测量结果来保证。如 A、B、C 三人同时测定某一个物理量,测定次数相同,其结果如图 1-2-2 所示。A 的测量结果精密度和准确度都高;B 的测量精密度虽高,但准确度低;C 的测量结果的精密度和准确度均低。

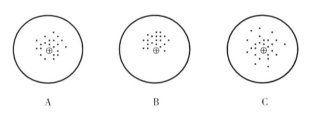

⊕—真值　·—测量值

图 1-2-2　A、B、C 三人的测量结果

1.2.4 偶然误差的统计规律和可疑值的舍弃

偶然误差符合正态分布规律,即正、负误差具有对称性,所以,只要测量次数足够多,在消除了系统误差和过失误差的前提下,测量值的算术平均值趋近于真值。

但是,一般测量次数不可能有无限多次,所以一般测量值的算术平均值也不等于真值。于是人们又常把测量值与算术平均值之差称为偏差,常与误差混用。如果以误差出现次数 N 对标准偏差的数值 σ 作图,得一对称曲线(如图 1-2-1)。统计结果表明测量结果的偏差大于 3σ 的概率不大于 0.3%。因此,根据小概率定理,凡误差大于 3σ 的点,均可以作为过失误差剔除。严格地说,测量达到 100 次以上时方可如此处理,可粗略地用于 15 次以上的测量。10~15 次时可用 2σ,若测量次数再少,应酌情递减。

例如,相同条件下对某系统温度测量 15 次,结果如表 1-2-1 所示。试问第 8 次的测量值是否应予剔除?

表 1-2-1　相同条件下对某系统温度测量 15 次的实验结果

i	x_i	d_i	d_i^2
1	20.42	0.02	0.004
2	20.43	0.03	9
3	20.40	0.00	0
4	20.43	0.03	9
5	20.42	0.02	4
6	20.43	0.03	9
7	20.39	-0.01	1
8	20.30	-0.10	100
9	20.40	0.00	0

表 1-2-1(续)

i	x_i	d_i	d_i^2
10	20.43	0.03	9
11	20.42	0.02	4
12	20.41	0.01	1
13	20.39	-0.01	1
14	20.39	-0.01	1
15	20.40	0.00	0
	$\bar{x}=20.40$		$\sum d_i^2 = 0.015\ 2$

由表中数据计算,得

$$3\sigma = 3 \times \sqrt{\frac{\sum d_i^2}{n-1}} = 3 \times \sqrt{\frac{0.015\ 2}{14}} = 3 \times 0.033 = 0.099$$

第 8 次测量值的偏差为

$$|d_8| = |x_8 - \bar{x}| = |20.30 - 20.40| = 0.100 > 0.099$$

所以第 8 次的测量值应予剔除。剔除后,得

$$\sum d_i^2 = 0.005\ 2$$

$$3\sigma = 3 \times \sqrt{\frac{0.005\ 2}{13}} = 3 \times 0.02 = 0.06$$

所剩 14 次的偏差均不超过 0.06,故不必再剔除。

1.2.5　间接测量结果的误差传递

大多数情况下,要通过对几个简单的物理量进行测量,通过函数关系的计算来获得所需的实验结果。各测量步骤所引起的测量误差必然通过计算传递到最后结果,而影响准确度。下面给出了误差传递的定量公式。通过间接测量结果误差的求算,可以知道哪个直接测量值的误差对间接测量结果影响最大,从而可以有针对性地提高测量仪器的精度,获得好的结果。

1. 间接测量结果的误差计算

一般而言,设有函数 $u = F(x,y)$,其中 x、y 为可以直接测量的量。则

$$\mathrm{d}u = \left(\frac{\partial u}{\partial x}\right)_y \mathrm{d}x + \left(\frac{\partial u}{\partial y}\right)_x \mathrm{d}y$$

此为误差传递的基本公式。若 Δu、Δx、Δy 为 u、x、y 的测量误差,且设它们足够小,可以代替 $\mathrm{d}u$、$\mathrm{d}x$、$\mathrm{d}y$,几种简单函数关系间接测量结果的绝对误差和相对误差如表 1-2-2 所示。

表 1-2-2 几种简单函数关系间接测量结果的绝对误差和相对误差

函数关系	绝对误差	相对误差
$u = x + y$	$\pm(\|\Delta x\| + \|\Delta y\|)$	$\pm\left(\dfrac{\|\Delta x\| + \|\Delta y\|}{x + y}\right)$
$u = x - y$	$\pm(\|\Delta x\| + \|\Delta y\|)$	$\pm\left(\dfrac{\|\Delta x\| + \|\Delta y\|}{x - y}\right)$
$u = xy$	$\pm(x\|\Delta y\| + y\|\Delta x\|)$	$\pm\left(\dfrac{\|\Delta x\|}{x} + \dfrac{\|\Delta y\|}{y}\right)$
$u = x/y$	$\pm\left(\dfrac{x\|\Delta y\| + y\|\Delta x\|}{y^2}\right)$	$\pm\left(\dfrac{\|\Delta x\|}{x} + \dfrac{\|\Delta y\|}{y}\right)$
$u = x^n$	$\pm(nx^{n-1}\Delta x)$	$\pm\left(n\dfrac{\Delta x}{x}\right)$
$u = \ln x$	$\pm\left(\dfrac{\Delta x}{x}\right)$	$\pm\left(\dfrac{x\Delta x}{x\ln x}\right)$

2. 间接测量结果的标准偏差计算

直接测量的实验数据假设为 x 和 y，与间接测量数据 u 的关系符合函数 $u = f(x, y)$，则函数 u 的标准偏差为

$$\sigma_u = \sqrt{\left(\dfrac{\partial u}{\partial x}\right)_y^2 \sigma_x^2 + \left(\dfrac{\partial u}{\partial y}\right)_x^2 \sigma_y^2}$$

几种函数关系间接测量结果的绝对标准偏差和相对标准偏差见表 1-2-3。

表 1-2-3 几种函数关系间接测量结果的绝对标准偏差和相对标准偏差

函数关系	绝对标准偏差	相对标准偏差
$u = x \pm y$	$\pm\sqrt{\sigma_x^2 + \sigma_y^2}$	$\pm\dfrac{1}{x \pm y}\sqrt{\sigma_x^2 + \sigma_y^2}$
$u = xy$	$\pm\sqrt{y^2\sigma_x^2 + x^2\sigma_y^2}$	$\pm\dfrac{1}{x \pm y}\sqrt{\sigma_x^2 + \sigma_y^2}$
$u = x/y$	$\pm\dfrac{1}{y}\sqrt{\sigma_x^2 + \dfrac{x^2}{y^2}\sigma_y^2}$	$\pm\sqrt{\dfrac{\sigma_x^2}{x^2} + \dfrac{\sigma_y^2}{y^2}}$
$u = x^n$	$\pm nx^{n-1}\sigma_x$	$\pm\dfrac{n}{x}\sigma_x$
$u = \ln x$	$\pm\dfrac{\sigma_x}{x}$	$\pm\dfrac{\sigma_x}{x\ln x}$

1.2.6 测量结果的有效数字

物理量的数值不仅能反映出量的大小、数据的可靠程度,而且还反映了仪器的精确程度等问题。当我们对一个测量的量进行记录时,所记数字的位数应与仪器的精密度相符合,即所记数字的最后一位为仪器最小刻度以内的估计值,称为可疑值,其他几位为准确值。这样一个数字称为有效数字,它的位数不可随意增减。例如,普通 50 mL 的滴定管,最小刻度为 0.1 mL,则记录 27.55 是合理的,记录 27.5 和 27.556 都是错误的,因为它们分别缩小和放大了仪器的精密度。为了方便地表达有效数字位数,一般用科学记数法记录数字,即用一个带小数的个位数乘以 10 的相应幂次表示。如 0.000 667 可写为 6.67×10^{-4},有效数字为三位;10 670 可写为 $1.067\ 0 \times 10^4$,有效数字是五位。用以表达小数点位置的零不计入有效数字位数。

在间接测量中,须通过一定公式将直接测量值进行运算,运算中对有效数字位数的取舍应遵循如下规则:

(1) 误差一般只取一位有效数字,最多两位。

(2) 有效数字的位数越多,数值的精确度也越高,相对误差越小;任何一物理量的数值,其有效数字的最后一位,在位数上应与误差的最后一位一致。

① 1.25 ± 0.01 m,三位有效数字,相对误差 0.8%。

② $1.250\ 0 \pm 0.000\ 1$ m,五位有效数字,相对误差 0.008%。

将 1.25 ± 0.01 m 三位有效数字写成 1.251 ± 0.01 m 或 1.3 ± 0.01 m 都不正确。

(3) 若第一位的数值等于或大于 8,则有效数字的总位数可多算一位。

如 9.23 虽然只有三位,但在运算时,可以看成四位。

(4) 运算中舍弃过多不定数字时,应用"4 舍 6 入,逢 5 尾留双"的法则。

如有下列两个数值:9.435 和 4.685,整化为三位数,根据上述法则,整化后的数值为 9.44 与 4.68。

(5) 在加减运算中,各数值小数点后所取的位数,以其中小数点后位数最少者为准。

如 $56.38 + 17.889 + 21.6 = 56.4 + 17.9 + 21.6 = 95.9$

(6) 在乘除运算中,各数保留的有效数字,应以其中有效数字最少者为准。

如式 $1.436 \times 0.020\ 568 \div 85$ 中 85 的有效数字最少,由于首位是 8,所以可以看成三位有效数字,其余两个数值,也应保留三位,最后结果也只保留三位有效数字。

如 $$\frac{1.44 \times 0.020\ 6}{85} = 3.49 \times 10^{-4}$$

(7) 在乘方或开方运算中,结果可多保留一位。

(8) 对数运算时,对数中的首数不是有效数字,对数尾数的位数应与各值的有效数字相当。

如
$$[H^+] = 7.6 \times 10^{-4}$$
$$pH = 3.12$$
$$K = 3.4 \times 10^9$$
$$\lg K = 9.35$$

(9) 计算公式中的一些常数 π、e、阿伏伽德罗常数、普朗克常数等,不受上述规则限制,其位数按实际需要取舍。

1.2.7 实验数据的表达

为了阐明和分析某些规律,需将实验数据归纳、处理,常用列表法、作图法和方程式法建立数值间的相互关系,现分述如下。

1. 列表法

做完实验后,所获得的大量数据,应尽可能整齐地、有规律地列表表达出来,使得全部数据能够一目了然,便于运算处理,容易检查而减少差错。

列表时应注意以下几点:

(1) 每一个表都应有完整的名称;

(2) 在表的每一行或每一列的第一栏,要详细写出名称、数量、单位;

(3) 表中的数据应化为最简单的形式,公共的乘方因子应在第一栏的名称下注明;

(4) 在每一行中数字排列要整齐,位数和小数点要对准;

(5) 原始数据可与处理的结果并列在一张表上,处理方法和运算公式在表下注明。

2. 作图法

利用图形表达实验结果有许多好处,首先它能直接显示出数据的特点,如极大、极小、转折点等;其次能够利用图形作切线、求函数的微商、求面积、外推值等;将数据进一步地进行处理,用处极为广泛。作图法要注意以下几点。

(1) 坐标纸和比例尺的选择

最常用的坐标纸是直角坐标纸,其他如对数坐标纸、半对数坐标纸和三角坐标纸有时也用到。

在用直角坐标纸作图时,以自变量为横轴,因变量为纵轴,横轴与纵轴的读数一般不一定从 0 开始,视具体情况而定。坐标轴上比例尺的选择极为重要。由于比例尺的改变,曲线形状也将跟着改变,若选择不当,会导致曲线的某些相当于极大、极小或转折点的特殊部分看不清楚。比例尺的选择应遵守下述规则:

① 要能表示出全部有效数字,以便作图法求出的物理量的准确度与测量的准确度相适应,为此,最好将测量误差较小的量取较大的比例尺;

② 图纸每小格所对应的数值应便于迅速、简便地读数,便于计算;

③ 在上述条件下,充分考虑利用图纸的全部面积,使全部布局匀称合理;

④ 若作的图形是直线,则比例尺的选择应使其斜率接近于 45。

(2) 画坐标轴

选定比例尺后,画上坐标轴,在轴旁注明该轴所代表变数的名称及单位。横轴读数自左至右,纵轴自下至上。

(3) 作代表点

将测得数值的各点绘于图上,在点的周围画上圆圈、方块或其他符号(在有些情况下其面积大小应近似地显示测量的准确度,例如,若测量的准确度很高,圆圈就画得小些,反之就大些)。在一张图纸上如有数组不同的测量值时,各组测量值的代表点应用不同符号表示,以示区别,并须在图上注明。

(4) 连曲线

作出各代表点,用曲线板或曲线尺作出尽可能接近于实验点的曲线,曲线应光滑均匀,细而清,曲线不必强求通过所有各点,但各点在曲线两旁的分布,在数量上应近似于相等,代表点和曲线间的距离表示测量的误差,曲线与代表点之间的距离应尽可能小,并且曲线两侧各代表点与曲线间距离之和亦应近似相等。

(5) 写图名

写上清楚而完整的图名以及坐标轴的比例尺。图上除图名、曲线、坐标轴及读数之外,一般不再写其他的字及作其他的辅助线,以免主要部分反而不清楚。数据亦不要写在图上,但在报告上应有相应的、完整的数据。

3. 方程式法

当一组实验数据用列表法或图形法表示后,进一步常需用数字方程式表示出来。因为经验公式不仅形式紧凑,也便于求微分、积分或插值。经验方程式是客观规律的一种近似描述,它是理论探讨的线索和根据。

对于一组实验数据,一般没有一个简单的方法可以直接获得理想的经验公式。通常是先将一组实验数据画图,根据经验和解析几何原理,或者参考有关文献,确定公式的形式。公式中最直观的为直线形式,因此凡在许可情况下,应尽可能使所得函数为一直线式。对指数方程 $y = be^{ax}$ 或 $y = bx^a$ 可取对数,即 $\ln y = ax + \ln b$ 或 $\ln y = a\ln x + \ln b$,这样,若以 $\ln y$ 对 x 或 $\ln y$ 对 $\ln x$ 作图,均可得直线而求出 a 和 b。

上面只是简单地提到怎样由作图求经验方程,除此之外还有其他几种方法,在这里就不叙述了。

1.3 用计算机处理实验数据和表达实验结果

随着科学技术的进步,特别是近年来信息科学技术的发展,使得信息技术在物理化学实验中得到越来越广泛的应用。在物理化学实验中,使用的智能化、数字化仪器设备越来越多,获得数据的方式发生了很大的变化,处理实验数据与表达实验结果的方法也相应发生了改变。在处理实验数据和表达实验结果时,计算机的使用越来越广泛。在物理化学实验课程中,特别是撰写实验报告时,经常需要用表格列出实验数据和实验结果,根据数据作出相应的图形、作直线求斜率和截距、绘制曲线求各点切线的斜率,等等。

计算机软件种类很多,并且不断升级,发展很快。在实验中和撰写实验报告时,可以利用的软件也比较多。下面举例介绍两种常用的工具软件(Excel 和 Origin)在基础物理化学实验数据处理与结果表达中的应用。

1.3.1 用 Excel 列表处理数据并作图

在液体饱和蒸气压测定实验中,直接测量了 8 个温度及对应的真空度。数据处理时,要计算蒸气压、$1/T$、$\ln p$,作 $\ln p - 1/T$ 图,拟合直线求斜率,计算平均摩尔汽化焓。用 Excel 处理数据及作图步骤如下:

(1) 启动 Excel,将大气压、8 个温度及对应的真空度数据填入表格的 A、B、C 列中,在

D2～D9格中输入公式计算蒸气压。例如先选定D2格,然后在函数栏(f_x)中输入函数"=A2-C2",回车即得D2值18.04,如图1-3-1所示;然后再次选定D2,把鼠标置于D2右下角,鼠标图标由白色空心十字变为黑色实心十字,按下鼠标左键,往下滑动鼠标至D9,放开鼠标左键,则所有的D列蒸气压数据就显示在对应的位置。用同样的方法,在E2格中输入公式"=1 000/(B2+273.15)"计算1 000/T,在F2格中输入公式"=LN(D2*1 000)"计算ln p,计算结果如图1-3-2所示。

	A	B	C	D	E	F
1	大气压/Kpa	温度/℃	真空度/kPa	蒸气压/kPa	[1/(T/K)]×1000	ln(p/Pa)
2	101.12	32.80	83.08	18.04		
3		36.80	79.00			
4		40.10	76.08			
5		44.90	70.48			
6		49.70	63.82			
7		54.40	56.10			
8		60.30	44.80			
9		66.00	31.80			

图1-3-1 在表格中输入原始数据及计算公式

	A	B	C	D	E	F
1	大气压/Kpa	温度/℃	真空度/kPa	蒸气压/kPa	[1/(T/K)]×1000	ln(p/Pa)
2	101.12	32.80	83.08	18.04	3.27	9.80
3		36.80	79.00	22.12	3.23	10.00
4		40.10	76.08	25.04	3.19	10.13
5		44.90	70.48	30.64	3.14	10.33
6		49.70	63.82	37.30	3.10	10.53
7		54.40	56.10	45.02	3.05	10.71
8		60.30	44.80	56.32	3.00	10.94
9		66.00	31.80	69.32	2.95	11.15

图1-3-2 在表格中输入原始数据及计算公式后所得结果

(2)选定需要作图的两列数据,横坐标在左,纵坐标在右,依次点击菜单栏中的"插入""图表",在"图表类型"中选择"XY散点图",并在"子图表类型"中选择"散点图",如图1-3-3所示,然后点击"完成"即得所需散点图,如图1-3-4所示。

图 1-3-3 用 Excel 作图图表类型的选择

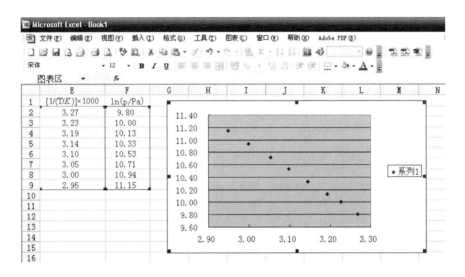

图 1-3-4 用 Excel 作图所得散点图

然后对所得散点图进行线性拟合,用左键点击选择图中数据点,右键弹出快捷菜单,选择"添加趋势线",在"类型"选项中选择"线性(L)",在"选项"中选择"显示公式",然后点击"确定"即得拟合直线,并给出了拟合直线的线性方程,如图1-3-5所示。

从拟合直线的线性方程中可获得直线的斜率及截距等信息。在 Excel 表中任选一空单元格,输入计算平均摩尔汽化焓的公式,即可获得平均摩尔汽化焓,如图1-3-6所示。

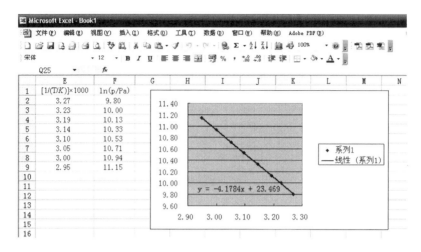

图1-3-5 用Excel作图对数据进行线性拟合

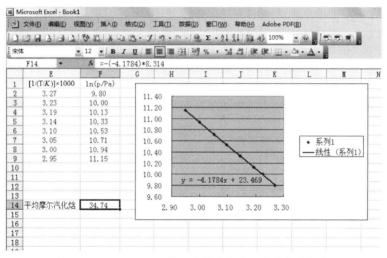

图1-3-6 用Excel作图由斜率求出平均摩尔汽化焓

最后将表格中的数据及图形拷贝到Word中打印即可,也可以对表格及图形作进一步编辑,以使其更加美观。

1.3.2 用Origin处理数据并作图

Origin是一个功能强大的数据处理及作图软件,作出的专业图形也比较规范。以下以Origin 7.0为例给出三个示例说明数据处理及作图步骤。

1. 用Origin 7.0处理饱和蒸气压测定实验数据及作图

(1)启动Origin 7.0程序,将大气压、实验所得沸点温度及对应的真空度(压力差)数据填入表格的A、B、C列中,然后输入公式计算D列(蒸气压/mmHg)的值,操作为左键点击选定D列,右键点击选择"Set Column Values",在弹出的对话框中输入计算公式"$p_{大气}$ - 压力差",本例为"767.65 - col(C)",如图1-3-7所示,点击"OK"完成D列值的设置。按此方法依次输入公式"1 000/(col(B) +273.15)""log(col(D))"设置E列和F列的值,所得结果如图 1-3-8 所示。

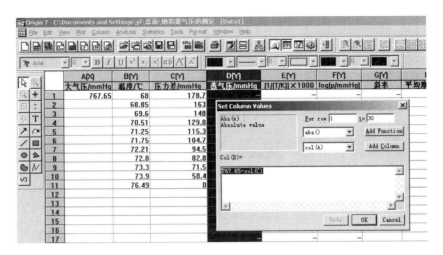

图 1-3-7　用 Origin 7.0 处理数据公式的设定

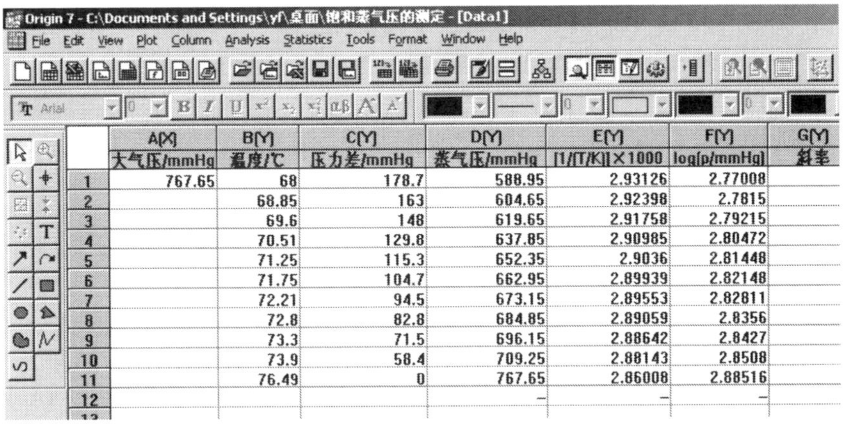

图 1-3-8　用 Origin 7.0 处理数据结果

（2）对上述所得数据作图：点击菜单栏中的"Plot"，然后选择"Scatter"，弹出如图 1-3-9 所示对话框，在列表中选择所需列为 X 或 Y，本例中以 E 列作为 X，即选中 E[Y]列，点击"<->X"键，如图 1-3-9 中箭头所示，F 列作为 Y，即选中 F[Y]列，点击"<->Y"键，然后点击"OK"即给出散点图，如图 1-3-10 所示。若要作多组散点图，可以在图 1-3-9 所示对话框中选定一组 X、Y 后点击"Add"，然后继续添加相应列为 X 和 Y 即可。作散点图的方法也可以是先直接将 E 列设置为 X，方法是选中 E 列，点击菜单栏中的"Column"→"Set as X"，即设为"E[X2]"，同时 F 列也变为"F[Y2]"，然后同时选中 E[X2]列和 F[Y2]列，点击菜单栏中的"Plot"，选择"Scatter"亦可得到如图 1-3-10 所示结果。

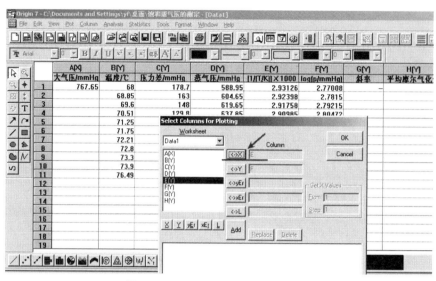

图 1-3-9　用 Origin 7.0 作图方法

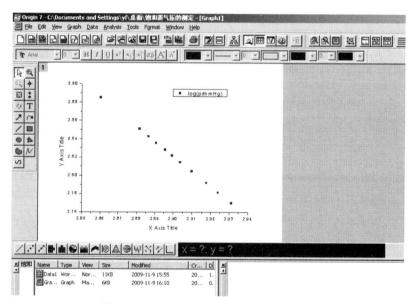

图 1-3-10　用 Origin 7.0 作散点图结果

然后对所得散点图进行线性拟合,方法是左键点击"Analysis",选择"Fit linear",即得拟合的直线,并在右下端窗口给出拟合后的线性方程,其斜率 B、截距 A 以及相关性系数 R 等信息,如图 1-3-11 所示。如果右下端窗口未显示相关信息,则点击菜单栏中"View"→"Results Log"即可显示。之后再回到 Data1 页面,在 G 列中设置计算平均摩尔汽化焓的公式,并将拟合得到的斜率代入即可计算出平均摩尔汽化焓。最后回到 Graph1 页面,调整好坐标轴刻度,横坐标为 $1\,000/T(K)$,纵坐标为 $\log p$ 或 $\ln p$。将所得结果直接用 Origin 7.0 软件打印或点击菜单栏中"Edit"→"Copy Page",然后粘贴到 Word 文档中打印。

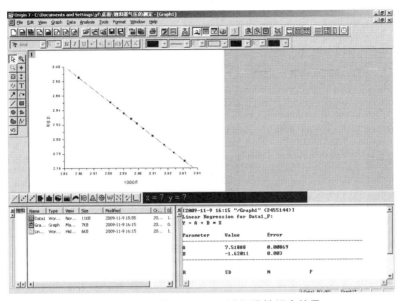

图 1-3-11　用 Origin 7.0 进行线性拟合结果

2.用 Origin 7.0 处理二元液系气液平衡相图实验数据并作图

(1)工作曲线的绘制

①启动 Origin 7.0 程序,将正丙醇含量、所测的两组折射率数据填入表格的 A、B、C 列中,然后左键点击选定 D 列,右键点击选择"Set Column Values",在弹出的对话框中输入计算公式"(col(B)+col(C))/2",如图 1-3-12 所示,点击"OK"完成 D 列值的设置,即求得两次折射率的平均值,如图 1-3-13 所示。

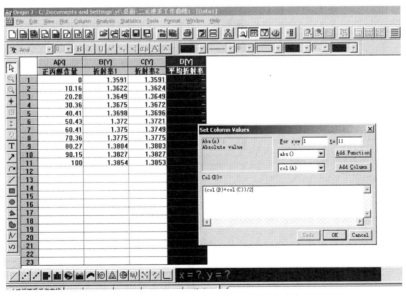

图 1-3-12　用 Origin 7.0 进行折射率数据处理

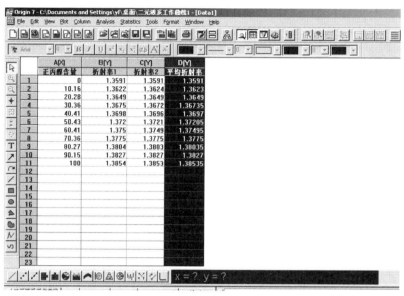

图1-3-13 用 Origin 7.0 进行折射率数据处理结果

②选中 D 列(Y),以 D 列对 A 列(X)作图,点击工具栏中的散点图快捷按钮,如图1-3-14箭头所指,即给出所需散点图,如图1-3-15所示。

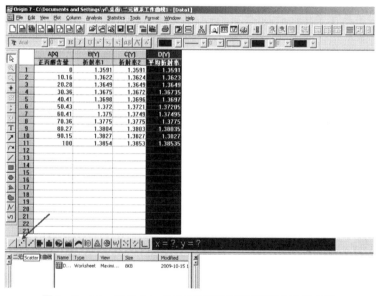

图1-3-14 用 Origin 7.0 作折射率对正丙醇含量散点图

③对所得散点进行拟合,方法是左键点击"Analysis",选择"Fit linear",即得拟合的直线,并在右下端窗口给出了拟合后的线性方程,其斜率 B、截距 A 以及相关性 R 等信息如图1-3-16所示,如果右下端窗口未显示相关信息,则点击菜单栏中"View"→"Results Log"即可显示,记录下所得线性方程。该方程即为折射率与浓度的关系,若实验测得任一正丙醇与乙醇混合体系的折射率即可通过该公式计算出其浓度,反之亦然。

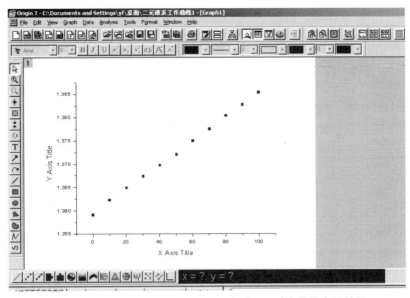

图 1-3-15　用 Origin 7.0 作折射率对正丙醇含量散点图结果

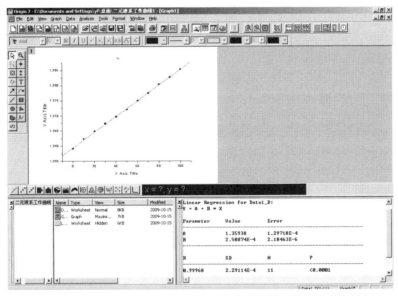

图 1-3-16　用 Origin 7.0 拟合后的直线及拟合出的线性方程

(2)二元液系相图的绘制

①启动 Origin 7.0 程序,将实验所得沸点、对应的液相折射率及气相折射率数据分别输入表格的 A、B、C 列中,根据上述线性方程,用"Set Column Values"设定 D 列和 E 列的值。本例中 D 列值设定公式为"(col(B) - 1.359 38)/0.000 258 874",E 列值设定公式为"(col(C) - 1.359 38)/0.000 258 874",即得不同沸点及对应的液相/气相中正丙醇的组成,如图 1-3-17 所示。

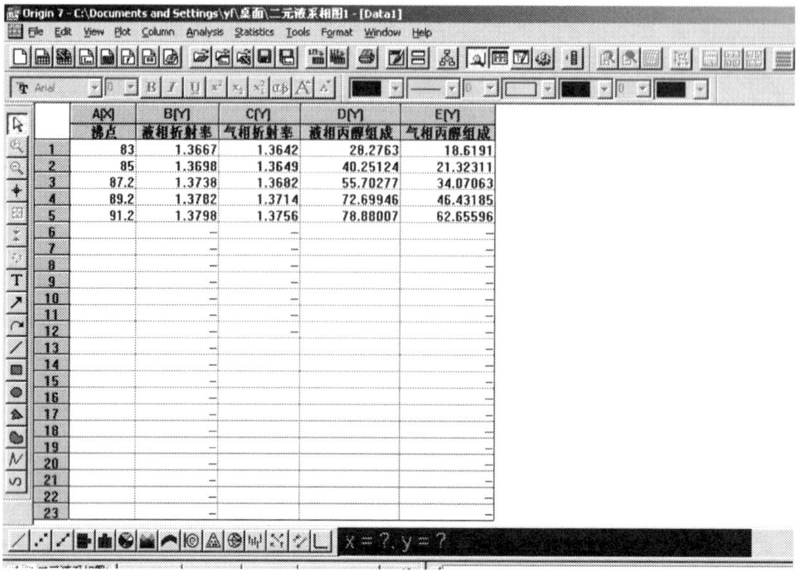

	A(X) 沸点	B(Y) 液相折射率	C(Y) 气相折射率	D(Y) 液相丙醇组成	E(Y) 气相丙醇组成
1	83	1.3667	1.3642	28.2763	18.6191
2	85	1.3698	1.3649	40.25124	21.32311
3	87.2	1.3738	1.3682	55.70277	34.07063
4	89.2	1.3782	1.3714	72.69946	46.43185
5	91.2	1.3798	1.3756	78.88007	62.65596

图 1-3-17　用 Origin 7.0 处理折射率和组成数据

②新建一个 Origin 工作表,选择菜单栏"Column"中的"Add New Columns"添加两列,并将 C 列设为 X,方法是先用鼠标左键选中 C 列,然后选择菜单栏"Column"中的"Set as X",然后在 A 列中输入上步所得液相正丙醇的组成,C 列中输入上步所得气相正丙醇的组成,在 B 列和 D 列中输入相应沸点,并将正丙醇组成为 0(即纯乙醇)时的沸点 78.5 ℃和正丙醇组成为 100(即纯正丙醇)时的沸点 97.4 ℃同时输入,结果如图 1-3-18 所示。

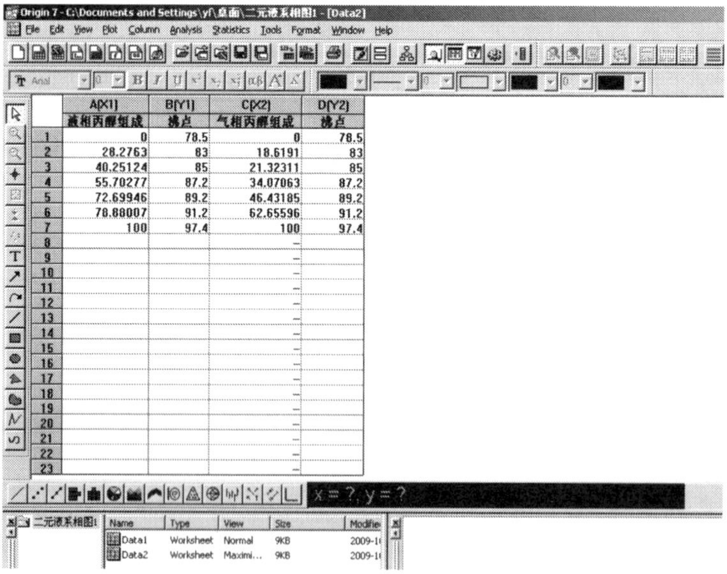

	A(X1) 液相丙醇组成	B(Y1) 沸点	C(X2) 气相丙醇组成	D(Y2) 沸点
1	0	78.5	0	78.5
2	28.2763	83	18.6191	83
3	40.25124	85	21.32311	85
4	55.70277	87.2	34.07063	87.2
5	72.69946	89.2	46.43185	89.2
6	78.88007	91.2	62.65596	91.2
7	100	97.4	100	97.4

图 1-3-18　用 Origin 7.0 处理折射率和组成数据结果

③选中图 1-3-18 所示 4 列数据,选择菜单栏"Plot"→"Special line/symbol"→"Double-Y"进行作图,结果如图 1-3-19 所示。点击图层 1,点击下端工具栏中的散点图快捷键将曲线 1 中的线去掉,同样方法点击图层 2,点击下端工具栏中的散点图快捷键将曲

线2中的线去掉,详细如图1-3-19中箭头所指,结果如图1-3-20所示。

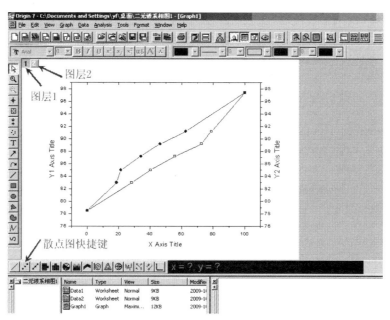

图1-3-19 用 Origin 7.0 作沸点-组成(气相及液相)图

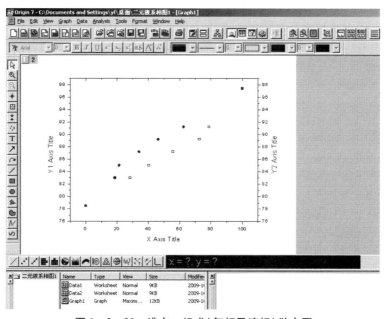

图1-3-20 沸点-组成(气相及液相)散点图

④对所得两条曲线散点进行拟合,方法是分别选择图层1和图层2,左键点击菜单栏中的"Analysis"并选择"Fit Gaussian"或"Fit Sigmoldal"进行拟合,调节好坐标轴将横坐标改成0~100,完成相图,结果如图1-3-21所示。

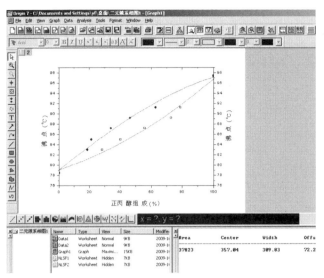

图 1-3-21 二元液系气液平衡相图

将所得工作曲线及相图结果直接用 Origin 7.0 软件打印或点击菜单栏中"Edit"→"Copy Page",然后粘贴到 Word 文档中打印。

3. 用 Origin 7.0 处理表面活性剂临界胶团浓度的测定实验数据并作图

①启动 Origin 7.0 程序,将溶液浓度及压差原始数据输入表格中的 A、B、C、D 列中,然后选中 E 列,右键点击选择"Set Column Values",在弹出的对话框中输入计算公式"(col(B)+col(C)+col(D))/3",点击"OK"完成 E 列值的设置,即得到平均压差。从附录中查出实验温度下水的表面张力,如 30 ℃时水的表面张力为 71.18 mN/m,然后选中 F 列,右键点击选择"Set Column Values",在弹出的对话框中输入计算公式"col(E) * 71.18/46.1",其中 46.1 为水的平均压差,点击"OK"完成 F 列值的设置,即得到各浓度下的表面张力,结果如图 1-3-22 所示。

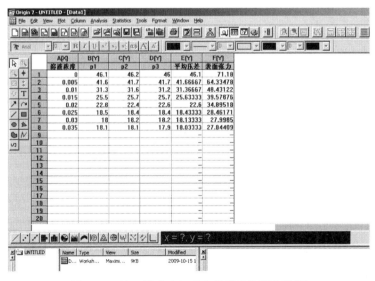

图 1-3-22 用 Origin 7.0 处理表面张力数据

②将溶液浓度和表面张力数据拷贝至一个新的 Origin 工作表中,结果如图 1 - 3 - 23 所示,选中 A、B 两列,采用"line + symbol"作点线图,结果如图 1 - 3 - 24 所示。

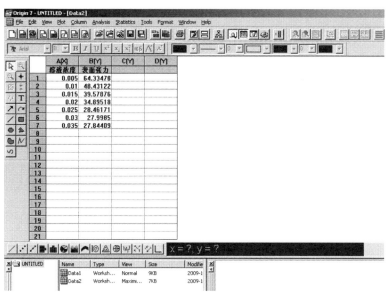

图 1 - 3 - 23　用 Origin 7.0 处理表面张力随浓度变化数据结果

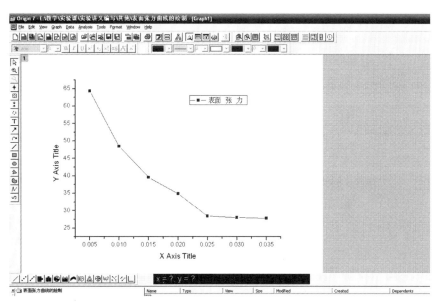

图 1 - 3 - 24　表面张力对浓度作图

③双击图中曲线,在弹出的"Plot Details""line"窗口中的"Connect"选项框选择"B - Spline",如图 1 - 3 - 25 所示,点击确定,即得表面张力曲线。

④从图上可知前四个点大体在一条直线上,后三个点在一条直线上,因此将它们分成两条直线作图,而这两条直线的交点即为临界胶束点,可由此求得临界胶束浓度。具体做法为:回到 Data2 界面,将后三组表面张力数据剪切到 C 列相应位置,如图 1 - 3 - 26 所示。

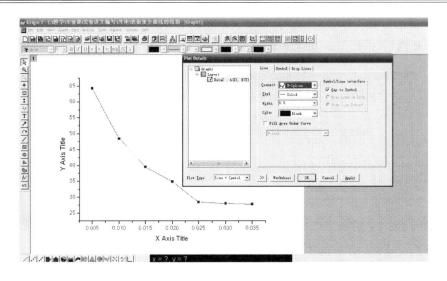

图 1-3-25 表面张力曲线

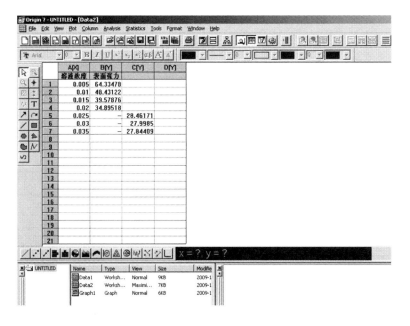

图 1-3-26 表面张力数据分组

⑤选中 A、B、C 三列如前所述作散点图,即得两组散点,如图 1-3-27 所示。

⑥对处于下降趋势的四个点进行线性拟合,方法是左键点击菜单栏中的"Analysis",选择"Fit linear",即得拟合的直线,并在右下端窗口给出了拟合后的线性方程,其斜率 B、截距 A 以及相关性 R 等信息如图 1-3-28 所示。如果右下端窗口未显示相关信息,则点击菜单栏中"View"→"Results Log"即可显示,将所得线性方程式记为方程式 1。

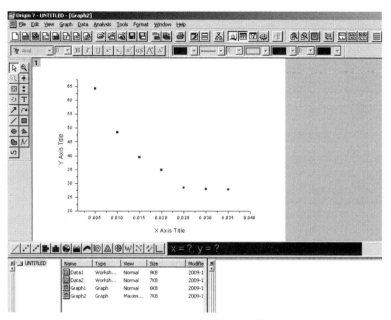

图1-3-27 表面张力曲线分组

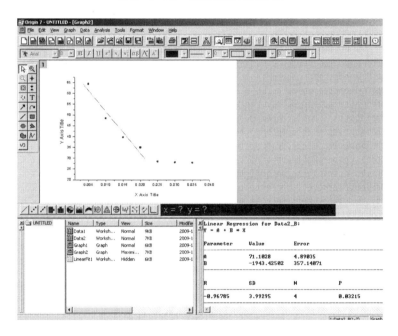

图1-3-28 对表面张力下降曲线进行线性拟合

⑦激活处于水平线上的三点,方法是在该三点的任意一点上点击鼠标右键,选择"Set as Active",如图1-3-29所示,然后按照第⑤步方法对该三点进行线性拟合,在右下端窗口给出了拟合后的线性方程,其斜率 B、截距 A 以及相关性 R 等信息如图1-3-30所示,将所得线性方程式记为方程式2。

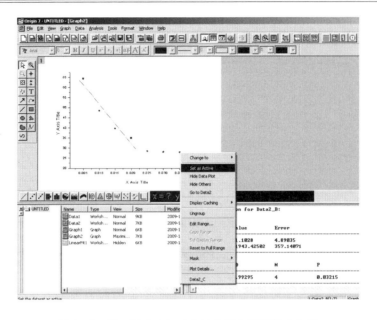

图 1-3-29　用 Origin 7.0 处理拟合表面张力曲线

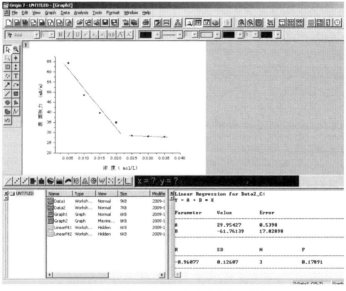

图 1-3-30　表面张力对浓度作图结果

⑧将所得线性方程式 1 和线性方程式 2 进行联立求解出两条直线的交点,所得交点的横坐标即为临界胶束浓度,交点纵坐标即为临界胶束浓度时所对应的表面张力。

⑨回到 Graph2 界面,调整两坐标轴标度及标注,横坐标标注为"浓度(mol/L)",纵坐标标注为"表面张力(mN/m)",结果如图 1-3-30 所示,最后将所得图片直接打印,或按前述方法拷贝到 Word 文档中进行打印。

Origin 7.0 作图功能比较强,可以编辑作出一些比较复杂的图形,这里仅给了三个简单的例子,以期起到抛砖引玉的作用,更加复杂图形的做法请同学们在实际应用中自己摸索。

第 2 章 基 础 实 验

实验 1 恒温水浴调节和黏度的测定

【实验目的】

1. 了解恒温水浴的构造及工作原理;
2. 掌握恒温水浴的调节及灵敏度的测定;
3. 掌握用乌氏黏度计测定液体黏度的方法;
4. 了解温度变化对黏度的影响。

【预习要求】

1. 了解恒温水浴的调节及灵敏度的测定方法;
2. 了解毛细管法测定黏度的原理。

【实验原理】

物质的物理化学性质如黏度、密度、蒸气压、表面张力、折光率、电导、电导率、透光率等都随温度而改变,要测定这些性质必须在恒温条件下进行。一些物理化学常数如平衡常数、化学反应速率常数等也与温度有关,这些常数的测定也需要恒温。因此,学会恒温水浴的使用对物理化学实验是非常必要的。

恒温水浴的形式很多,区别主要在于控制温度的高低及控制温度的精度。图 2-1-1 的装置是本实验所用的恒温水浴(HK-1D 型恒温水浴,南京大学应用物理研究所制造),它适合在高于室温但低于水的沸点温度范围内工作。

恒温水浴的性能指标可通过测定恒温后的恒温水浴的最高和最低温度来计算灵敏度。灵敏度 t_E 与最高温度 t_1、最低温度 t_2 的关系式为 $t_E = \pm \dfrac{t_1 - t_2}{2}$。恒温后恒温水浴的温度仍有一定的波动范围,可用恒温水浴的灵敏度来表征,t_E 越小,恒温水浴的性能越好。若将温度恒定在 30.00 ℃,则恒温水浴的温度记作 30.00 ℃ $\pm t_E$。

图 2-1-1 HK-1D 型恒温水浴

液体黏度的大小一般用黏度系数(简称黏度,η)表示。测定黏度的方法主要有毛细管法、转筒法和落球法。在测定高聚物分子的特性黏度时,以毛细管流出法的(Ubbelohde)黏度计最为方便。毛细管黏度计常用的有乌氏(Ubbelohde)黏度计和奥氏(Ostwald)黏度计,如图2-1-2所示。奥氏黏度计要求试样的体积必须每次都相同,操作过程中由于黏度计位置倾斜所导致的流出时间的误差也很大。若液体在毛细管黏度计中,因重力作用流出时,可通过泊肃叶(Poiseuille)公式计算黏度系数:

$$\eta = \frac{\pi p r^4 t}{8VL} \quad (2-1-1)$$

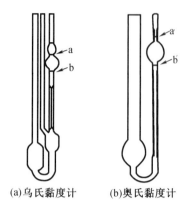

(a)乌氏黏度计　(b)奥氏黏度计

图 2-1-2　乌氏黏度计和奥氏黏度计

式中,η 为液体的黏度,在 C·G·S 制中黏度单位为泊(P),在 SI 中,黏度单位为帕·秒(Pa·s),1 P = 0.1 Pa·s;L 为毛细管的长度,m;r 为毛细管的半径,m;t 为流出的时间,s;V 为流经毛细管的液体体积,m³;p 为毛细管两端的压力差,N。由于 r、L、p 等数值不易测准,故按式(2-1-1)由实验来测定液体的绝对黏度是项很困难的工作,但测定液体对标准液体(如水)的相对黏度则是简单和适用的。在已知标准液体的绝对黏度时,也可算出被测液体的绝对黏度。

设两种液体在本身所受重力作用下分别流经同一毛细管,且流出的体积相等,则

$$\eta_1 = \frac{\pi r^4 p_1 t_1}{8VL} \quad (2-1-2)$$

$$\eta_2 = \frac{\pi r^4 p_2 t_2}{8VL} \quad (2-1-3)$$

从而

$$\frac{\eta_1}{\eta_2} = \frac{p_1 t_1}{p_2 t_2} \quad (2-1-4)$$

式中,$p = hg\rho$,h 为流过毛细管液体的平均液柱高度;ρ 为液体密度;g 为重力加速度。

如果每次取用试样的体积一定,则可保持 h 在实验中情况相同。则

$$\frac{\eta_1}{\eta_2} = \frac{\rho_1 t_1}{\rho_2 t_2} \quad (2-1-5)$$

已知标准液体的黏度,则被测液体的黏度可按式(2-1-5)算得。

【仪器与试剂】

HK-1D 型恒温水浴,乌氏黏度计 1 支,20 mL 移液管 2 支,秒表 1 块,乙醇,蒸馏水。

【实验步骤】

1. 接通电源,打开仪器电源开关,开动搅拌机,将温度"设定"调节至 30.00 ℃,然后再将开关打到"测量"挡,其他温度调节方法相同。这时红色指示灯亮,显示加热器在工作。

2.待恒温水浴达到指定温度,温度稳定 10 min 后,观察并记录红灯熄灭后仪器电子屏显示的最高示值及红灯出现后的最低示值,此后每隔 2 min 记录一次最高示值和最低示值,连续观察至此最高和最低示值的三次平均值与指定温度相差不超过 0.10 ℃为止。

3.在实验前应先用洗液浸泡黏度计约 30 min,然后用自来水冲洗,再用蒸馏水淋洗,烘干备用。

4.用移液管取 20 mL 待测液放入黏度计里,将黏度计垂直浸入恒温水浴中(确保刻度线 a 在水面以下),待内外温度一致后(一般要 15 min 以上),用洗耳球吸起液体超过黏度计上刻度(a 点),然后放开洗耳球,用秒表记录液面自上刻度(a 点)到下刻度(b 点)所经历的时间。再吸起液体,重复测试至少三次,取其平均值。

5.调节恒温水浴温度到 40 ℃,按上述第 4 步方法测定液体流经刻度线 a 到刻度线 b 的时间。

6.测试完成后,倾出黏度计中的液体,用蒸馏水洗净黏度计。

【实验记录及数据处理】

室温:_____℃　气压:_____ kPa　　恒温水浴设定温度:_____℃
30 ℃时水的密度:_____　　　　　　40 ℃时水的密度:_____
30 ℃时乙醇的密度:_____　　　　　40 ℃时乙醇的密度:_____

1.将所测的数据和结果按表 2-1-1 和表 2-1-2 形式列出。

表 2-1-1　恒温水浴灵敏度实验数据

次数	观测温度		
	最高温度	最低温度	灵敏度
1			
2			
3			
4			
5			
6			
7			
平均值			

表2-1-2 液体黏度的测定

时间/s	乙醇		水	
	30 ℃	40 ℃	30 ℃	40 ℃
t_1				
t_2				
t_3				
t_4				
t_5				
平均时间				
黏度				

2. 从附录中查出水和乙醇的密度以及纯水的绝对黏度,计算乙醇的黏度,并与手册数据比较,求百分误差。

【讨论与说明】

1. 用过的乙醇倒入回收瓶中。

2. 温度对液体黏度影响很大,故本实验必须把恒温水浴准确调为30.00 ℃。实验时往上吸提液体样品一定注意不要超出恒温水浴的液面。如果两次测量相差较大,很可能是液体样品尚未恒温所导致。

3. 实验顺序为先测乙醇后测水,可节省时间。这是因为测水之前可以用较多的蒸馏水冲洗若干次后再进行,相反,若先测水,换测乙醇前需将黏度计烘干,多花费时间。

4. 泊肃叶公式 $\eta = \pi p r^4 t / 8VL$,其中 $p = \rho h g$ 是压力差、推动力,每次测定时必须注意黏度计在恒温水浴中要垂直放置,否则每次液柱垂直高度 h 不同,测定结果会带进误差。

5. 移液管为专用移液管,注意标签,不要混用。

6. 用移液管加入乙醇或水时应该从黏度计的大口加入。

【思考题】

1. 哪些因素会影响恒温水浴的工作质量?
2. 测定黏度的方法有哪些?
3. 为什么用奥氏黏度计时,加入标准物及被测物的体积应相同,而用乌氏黏度计测黏度时则没有该要求?
4. 为什么测定黏度时要保持温度恒定?
5. 黏度计放在恒温水浴内为什么必须垂直?

【相关阅读】

液体黏度及其测量方法

黏度分为动力黏度和运动黏度,一般将动力黏度简称为黏度。

流体流动时流层间存在着速度差和运动逐层传递。当相邻流层间存在速度差时,快速流层力图加快慢速流层,而慢速流层则力图减慢快速流层。这种相互作用随着流层间速度差的增加而加剧。流体所具有的这种特性称为黏性,流层间的这种相互作用力称为内摩擦力或黏性(滞)力。(动力)黏度 η 是用来表示流体黏性程度的物理量,被定义为 $v_x = 0$ 的稳定层流中剪切应力 τ_{xz} 与剪切速率 $\dfrac{\mathrm{d}v_x}{\mathrm{d}z}$ 的比值,即

$$\tau_{xz} = \eta \frac{\mathrm{d}v_x}{\mathrm{d}z}$$

动力黏度的单位是帕斯卡·秒,记作 Pa·s,1 Pa·s = 1 (N/m²)·s = 1 kg·m⁻¹·s⁻¹。

实际工作中常常直接测量运动黏度 ν,其定义为(动力)黏度 η 与流体密度 ρ 之比,即

$$\nu = \frac{\eta}{\rho}$$

运动黏度的单位是平方米每秒,m²/s,具体工作中也用 mm²/s。

测定黏度的方法有下列几种。

旋转法:在两同轴圆筒间充以待测液体,当筒匀速转动时,可由测定内筒所受的黏滞力矩求得黏度。

落球法:如果一小球在黏滞液体中铅直下落,由于附着于球面的液层与周围其他液层之间存在着相对运动,因此小球受到黏滞阻力,它的大小与落球速度有关。测出落球的速度后可以计算出液体黏滞系数,这种方法一般用来测量黏度较大的液体,并要求液体有一定的透明度。

毛细管法:通过测定在恒定的压强差作用下,流经一毛细管的液体流量来计算黏度。

其他方法,如振动法、平板法、流出杯法等。

1. 旋转法测定液体的黏度

对牛顿流体,由牛顿内摩擦定律可知平面层流时流层间的内摩擦力等于表面积 s、黏度 η 和速度梯度 $\dfrac{\mathrm{d}v}{\mathrm{d}l}$ 的乘积,即

$$F = -\eta s \frac{\mathrm{d}v}{\mathrm{d}l} \qquad (2-1-6)$$

当液体产生稳定旋转时,对于高度为 L 的环状薄层,半径为 r 的表面所受的内摩擦力在柱坐标系中沿切线方向,其大小为

$$F_\phi = -\eta s\left(-r\frac{\mathrm{d}\omega}{\mathrm{d}r}\right) = -2\pi\eta L r\left(r\frac{\mathrm{d}\omega}{\mathrm{d}r}\right) \qquad (2-1-7)$$

式中的负号是因角速度 ω 沿径向递减之故。稳态旋转时半径为 r 的柱面所受的力矩为常数,设其值为 M_1,可得

$$M_1 = rF_\phi = -2\pi\eta L r^3 \frac{\mathrm{d}\omega}{\mathrm{d}r} \qquad (2-1-8)$$

利用边界条件 $\omega|_{r=R_2} = \omega_0$ 和 $\omega|_{r=R_1} = 0$ 可得 $\omega = \dfrac{M_1}{4\pi\eta L}\cdot\dfrac{1}{r^2} + c$;再利用边界条件可得

$\omega_0 = \dfrac{M_1}{4\pi\eta L} \cdot \dfrac{1}{R_2^2} + c$ 和 $\dfrac{M_1}{4\pi\eta L} \cdot \dfrac{1}{R_1^2} + c = 0$，最后可得 $\eta = \dfrac{R_2^2 - R_1^2}{4\pi L R_1^2 R_2^2} \cdot \dfrac{M_1}{\omega_0}$，此即 Couette-Margules 公式。

当用旋转法测量时，同步电机以稳定的速度旋转，连接刻度圆盘，再通过游丝和转轴带动转子旋转（图2-1-3）。如果转子未受到液体的阻力，则游丝、指针与刻度圆盘同速旋转，指针在刻度盘上指出的读数为"0"。反之，如果转子受到液体的黏滞阻力，则游丝产生扭矩，与黏滞阻力抗衡最后达到平衡，这时与游丝连接的指针在刻度圆盘上指示一定的读数（即游丝的扭转角），将读数乘以特定的系数即得到液体的黏度。

2. 用落球法测定液体的黏度

当金属小圆球在黏性液体中下落时，它受到三个铅直方向的力：小球的重力 $\rho g V$（V 是小球体积，ρ 是小球密度）、液体作用于小球的浮力 $\rho_0 g V$（ρ_0 是液体密度）和黏滞力 f（其方向与小球运动方向相反）。实验装置如图2-1-4所示。如果液体是无限深广的，而且小球的半径 r 和下落速度 v 均较小，则有

$$f = 6\pi\eta v r \tag{2-1-9}$$

上式称为斯托克斯公式，其中 η 是液体的黏度，v 是小球下落速度，r 是小球的半径。

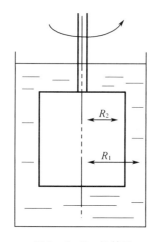

图2-1-3　旋转法

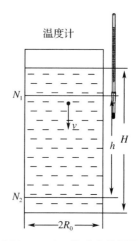

图2-1-4　落球法装置图

小球开始下落时，由于速度尚小，所以阻力也不大；但下落速度增大后，阻力也随之增大。最后，上述三个力达到平衡，即

$$\rho V g = \rho_0 V g + 6\pi\eta v r \tag{2-1-10}$$

于是，小球做匀速直线运动。由式(2-1-10)可得

$$\eta = \dfrac{(\rho - \rho_0)Vg}{6\pi v r} = \dfrac{2}{9}\dfrac{(\rho - \rho_0)g r^2}{v} \tag{2-1-11}$$

如已知 r、ρ、ρ_0 和 v 等值，即能计算 η。

Experiment 1　Determination of Liquid Viscosity

【Purpose of the Experiment】

1. To understand the concept of viscosity.
2. To learn how to determine the liquid viscosity by Ostwald viscometer.

【Principle】

Viscosity is an important property of fluid. It reflects the size of the generated shear stress. The shear stress of a differential volume unit is proportional to the perpendicular direction of fluid velocity gradient when fluid flows, this type of fluid is called Newtonian liquid. The shear stress obeys the formula (2-1-1)

$$F = -\eta A \mathrm{d}u/\mathrm{d}y \quad (2-1-1)$$

Where negative sign indicates the direction of the shear stress and flow is in the opposite direction. Where the scale factor η is called the absolute viscosity (Referred to as the viscosity).

This experimental determination of the viscosity is the type of Newtonian liquid. There are three kinds of methods to measure the viscosity of a liquid:

(1) To measure the time required for the liquid to flow through a capillary tube with a capillary viscosimeter;

(2) To measure the falling speed of a ball in the liquid with a ball viscosimeter;

(3) To measure the relative rotation between the liquid and a coaxial cylinder with a rotation viscosimeter.

Capillary method is different according to the viscosimeter. We use Ostwald viscometer as shown in Figure 2-1-1. When the liquid flows through the capillary viscosimeter because of gravitation, it obeys the Poiseuille law:

$$\eta = \frac{\pi r^4 g h \rho t}{8Vl} \quad (2-1-2)$$

Where r is the capillary radius, l is the capillary length, g is the gravity acceleration, h is the average height of liquid column passing through the capillary tube, ρ is the liquid density, V is the volume of liquid passing through the capillary tube, t is the flow time. The value of r and V and l is definite for a given viscosimeter. The value of h is definite when the viscosimeter is given and the volume of liquid in the viscosimeter is invariable. Therefore, when the liquid volume joined into the viscosity is invariable, the Poiseuille equation can be written as:

$$\eta = K\rho t \quad (2-1-3)$$

$$K = \frac{\pi r^4 g h}{8Vl} \quad (2-1-4)$$

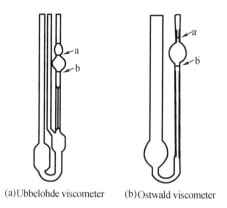

(a) Ubbelohde viscometer (b) Ostwald viscometer

Figure 2 – 1 – 1　Ubbelohde viscometer and Ostwald viscometer

Where K is the viscosimeter constant. Calculate constant K of the capillary viscosimeter by measuring the density and flow time of the liquid where the viscosity is known. In this experiment, the viscosity and density of distilled water at 30.0 ℃ are known as 0.797 5 centipoise and 0.995 6 g/mL. The density of alcohol at 30.0 ℃ is known as 0.780 9 g/mL. The viscosity of alcohol at 30.0 ℃ is be calculated as the formula (2 – 1 – 5)

$$\frac{\eta_{\text{alcohol}}}{\eta_{\text{water}}} = \frac{d_{\text{alcohol}} t_{\text{alcohol}}}{d_{\text{water}} t_{\text{water}}} \qquad (2-1-5)$$

【Apparatus and Reagents】

Thermostatic water bath; Ostwald viscosimeter; 10 mL pipette; aurilave; second chronograph; alcohol; distilled water.

【Procedures】

1. Install the equipments and set temperature for the temperature control unit. Set the thermostatic water bath's temperature to a specified temperature (30.0 ℃). Turn on the power of the thermostatic water bath and stirrer.

2. Measure the liquid viscosity.

Clean the viscosimeter and pipet 10.0 mL alcohol carefully into the viscosimeter. Clamp the viscosimeter vertically on an iron stand, and place it in the thermostatic water bath at 30.0 ℃, and make water surface of water bath higher than line a. After 10 min constant temperature, suck up the solution by the rubber suction bulb above line a, then let the solution fall along the capillary tube naturally. Start the stopwatch exactly as the liquid meniscus passes line a, and stop it just as the meniscus passes line b. Record the collapsed time for the liquid to fall from a to b. Repeat this measurement for three times.

Take the viscosimeter out and clean it with distilled water. Measure the t value through distilled water with the same amount of alcohol by the above method.

【Notes】

1. Stirring speed should be moderate and make sure the medium temperature is even.
2. Operation should be especially careful not to damage the viscosimeter.
3. The viscosimeter must be installed vertically.

【Data Analysis】

1. Calculate the viscosity of the liquid sample in the measured temperature, and compare it with literature values.

【Post Lab Questions】

1. Why the viscosimeter must be installed vertically? Through which physical parameters it affect the viscosity measurement in the Poiseuille equation?
2. Why the solution injected into the viscosimeter must be the same volume of standard solution?

实验 2 二元液系的气液平衡相图

【实验目的】

1. 绘制正丙醇 – 乙醇双液系的气 – 液平衡相图,了解相图和相律的基本概念;
2. 掌握测定双组分液体沸点的方法;
3. 掌握用折射率确定二元液体组成的方法。

【预习要求】

1. 理解绘制双液系相图的基本原理;
2. 了解阿贝折射仪的使用方法;
3. 了解本实验中的注意事项和判断气、液两相是否已达平衡的方法。

【实验原理】

1. 气 – 液相图

两种液态物质混合而成的二组分体系称为双液系。两个组分若能按任意比例互相溶解,称为完全互溶双液系。液体的沸点是指液体的蒸气压与外界压力相等时的温度。在一定的外压下,纯液体的沸点有其确定值。双液系的沸点不仅与外压有关,而且还与两种液体的相对含量有关。根据相律得

$$自由度 = 组分数 - 相数 + 2$$

因此，一个气-液共存的二组分体系的自由度为2，只要任意再确定一个变量，整个体系的存在状态就可以用二维图形来描述。例如，在一定温度下，可以画出体系的压力 p 和组分 x 的关系图；如体系的压力确定，则可作温度 T 对 x 的关系图。这种关系图就是相图。在 $T-x$ 相图上，还有温度、液相组成和气相组成三个变量，但只有一个自由度，一旦设定某个变量，则其他两个变量必有相应的确定值。图2-2-1(a)以正丙醇-乙醇体系为例表明，温度 T_1 这一水平线指出了在此温度时处于平衡的液相组分 x 和气相组分 y 的相应值。

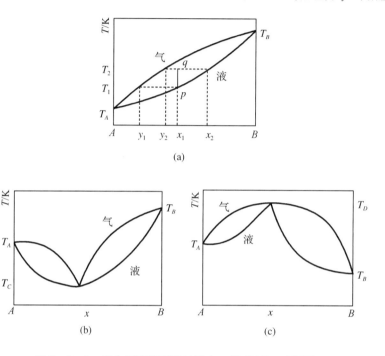

图 2-2-1 完全互溶双液系的沸点-组成（$T-x$）相图

正丙醇-乙醇这一双液系基本上接近于理想溶液。然而绝大多数实际体系与拉乌尔（Raoult）定律有一定偏差。偏差不大时，温度-组分相图与图2-2-1(a)相似，溶液的沸点仍介于两纯物质的沸点之间。但是，有些体系的偏差很大，以至其相图将出现极值。正偏差很大的体系在 $T-x$ 图上呈现极小值（图2-2-1(b)），负偏差很大时则会有极大值（图2-2-1(c)）。这样的极值称为恒沸点，其气、液两相的组成相同。例如，H_2O-HCl 体系的最高恒沸点在标准大气压时为 108.5 ℃，恒沸物的组成含 HCl 20.242%。

通常，测定一系列不同配比溶液的沸点及气、液两相的组成，就可绘制气-液体系的相图。下面以图2-2-1(a)为例，简单说明绘制沸点-组成相图的原理。加热总组成为 x_1 的溶液，体系的温度上升，达到液相线上的 p 点时溶液开始沸腾，组成为 y_1 的气相开始生成，但气相量很少（趋于0），x_1、y_1 两点代表达到平衡时液、气两相组成。继续加热，气相量逐渐增多，沸点继续上升，气、液二相组成分别在气相线和液相线上变化，当达某温度（如 q 点）并维持温度不变时，则 x_2、y_2 为该温度下液、气两相组成。分别取出气、液两相的样品，分析其组成，得到该温度下气、液两相平衡时各相的组成。改变溶液总组成，得到另一温度下，气、液两相平衡时各相的组成。测得溶液若干总组成下的气液平衡温度及气、液相组成，分别将气相点用线连接即为气相线,将液相点用线连接即为液相线,即得到沸点-组成相图。

2.沸点测定仪

本实验所用沸点测定仪如图2-2-2所示。

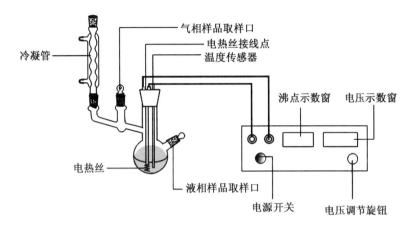

图2-2-2 沸点测定仪

3.组成分析

溶液的组成用折射率法分析。折射率法基于如下原理:在一定温度下纯物质各有一定折射率。如二液体形成混合物,则其折射率随组成的不同而变。因此,可在定温下测定若干组已知组成(x)溶液的折射率(n),绘出 $n-x$ 等温线(即工作曲线)。对于未知组成的样品,取出各样品后,测出该温度(测标准溶液 n 时所用温度)下的 n,便可从 $n-x$ 等温线查出其相应组成。因为乙醇和正丙醇的折射率相差很大,且折射法所需量较少,对本实验较适用。测折射率常用仪器为阿贝折射仪,其结构、原理及使用方法请参考本书第5章。

【仪器与试剂】

仪器:沸点测定仪1套,小滴瓶18个,移液管2支,玻璃滴管4支,WAY-2S型阿贝折射仪1台,超级恒温水浴1台。

试剂:乙醇(沸点:78.5 ℃),正丙醇(沸点:97.4 ℃)。

【实验步骤】

1.工作曲线绘制

(1)按表2-2-1配制各种不同浓度的乙醇-正丙醇标准溶液(注意,处理数据时应将体积换算成质量),分别放入带有编号的小滴瓶中,用阿贝折射仪测定其折射率,分别重复测量三次,取平均值。

(2)调节超级恒温水浴温度,使阿贝折射仪上的温度计读数保持在某一定值。分别测定上述溶液以及乙醇和正丙醇的折射率。

(3)绘制该温度下的折光率-组成工作曲线。

2. 安装沸点测定仪

根据图 2-2-2 所示,将沸点测定仪安装好,检查接口处连接是否紧密,电热丝接线是否良好。

3. 取样并测定

(1)打开"液相样品取样口"的塞子,用移液管加入 12 mL 乙醇、8 mL 丙醇,然后盖上玻璃塞,冷凝管通入冷水,打开沸点测定仪电源开关。

(2)调节电压至 15 V,一段时间后液体开始沸腾,维持回流状态 5 min 左右,然后从"沸点示数窗"处读取、记录液体沸点。

(3)调节电压至零,使体系冷却 1 min,然后用玻璃滴管从"气相样品取样口"取冷凝液,取完后,塞上塞子,用阿贝折射仪测定样品折射率,并记录。

(4)用玻璃管从"液相样品取样口"取液相样品,取完后,塞上塞子,用阿贝折射仪测定其折射率,并记录。

(5)打开"液相样品取样口"的塞子,用移液管加入 8 mL 丙醇,盖上玻璃塞,重复(2)~(4)步操作。

(6)重复步骤(5)四次。

【实验记录及数据处理】

1. 将乙醇-正丙醇标准溶液的折射率和组成填于表 2-2-1,并以折射率为纵轴、组成为横轴绘制工作曲线。

2. 利用上述工作曲线,由气相及液相样品的折射率,求出其相应的组成,填于表 2-2-2。

3. 以沸点为纵轴、气相及液相样品的组成为横轴绘制乙醇-正丙醇的沸点-组成相图。

室温:_____℃　　气压:_____ kPa

表 2-2-1　乙醇-正丙醇标准混合物折射率的测定

编号		1#	2#	3#	4#	5#	6#	7#	8#	9#	10#	11#
混合液 /mL	正丙醇	0	1	1	3	2	2	3	7	4	9	2
	乙　醇	2	9	4	7	3	2	2	3	1	1	0
正丙醇的质量分数/%												
折射率	1											
	2											
	3											
	平均											

表 2-2-2　乙醇-正丙醇二元液系实验数据

	编号		1#	2#	3#	4#	5#	6#	7#
	沸点/℃								
液相	折射率	1							
		2							
		3							
		平均							
	组成(正丙醇质量分数/%)								
气相	折射率	1							
		2							
		3							
		平均							
	组成(正丙醇质量分数/%)								

【讨论与说明】

阿贝折射仪的使用步骤如下：

(1) 打开棱镜,用擦镜纸将镜面擦拭干净后,在镜面上滴少量待测液体,并使其铺满整个镜面,关上棱镜(注意:滴加液体时滴管不要碰到棱镜,以防损坏棱镜)。

(2) 调节反射镜使入射光线达到最强,然后转动棱镜使目镜半明半暗,分界线位于十字线的交叉点,这时按下读数键可从液晶屏上读出液体的折射率。

(3) 如出现彩色光带,调节消色补偿器,使彩色光带消失,阴暗界面清晰。

(4) 测完之后,打开棱镜并用丙酮洗净镜面,也可用吸耳球吹干镜面,实验结束后,除必须使镜面清洁外,尚需夹上两层擦镜纸才能扭紧两棱镜的闭合螺丝,以防镜面受损。

(5) 其他需说明的问题:阿贝折射仪在使用前,必须先经标尺零点的校正,可用已知折射率的标准液体(如纯水的 $n_D^{20} = 1.3325$),亦可用每台折射仪中附有已知折射率的"玻块"来校正。可用 a-溴萘将"玻块"光的一面黏附在折射棱镜上,不要合上棱镜,打开棱镜背后小窗使光线由此射入,用上述方法进行测定,如果测得值和此"玻块"的折射率有区别,旋动镜筒上的校正螺丝进行调整。测折射率时,对于混合组分的测试要迅速,以免低沸点组分挥发,从而造成折射率数值测定的误差。由于折射率与温度有关,因此必须要使测定标准液与测定气液相样品折射率时的温度相同,这可以由折射仪的恒温来实现。腐蚀性液体如强酸、强碱和氟化物的测定不得使用阿贝折射仪。此外,在操作时还应注意:吸管绝对不允许碰到棱镜,以免划伤棱镜或万一碰碎吸管,留下玻璃碎片将仪器损坏。除专用的擦镜纸外,任何物品绝对不允许接触棱镜。每次测定完毕可将棱镜打开,使残留溶液挥发至干,或用吸耳球将棱镜表面吹干,必要时才用擦镜纸擦拭一下。

【思考题】

1. 溶液的折射率与哪些因素有关?
2. 作乙醇－正丙醇标准溶液的折射率－组成工作曲线的目的是什么?
3. 纯乙醇和正丙醇的沸点应如何确定?
4. 加入蒸馏瓶中的乙醇和正丙醇是否应该精确计量?
5. 整个实验为什么要防止任何方式引进水分?
6. 实验的误差来自何处?

【相关阅读】

相图的绘制

相图是表达多相体系的状态如何随温度、压力、组成等强度性质的变化而变化的几何图形,是根据实验数据在相律的指导下绘制的。实验测定相图的方法主要有三种。

1. 蒸馏法

蒸馏法一般用于气－液平衡系统(如本实验)相图的测定。该方法利用回流及分析的方法绘制相图,取不同组成的溶液在沸点测定仪中回流,测定其沸点及气、液相组成。沸点数据可直接由温度计读取,气、液相组成利用组成与折射率的关系,使用阿贝折射仪间接测得。

2. 热分析法(又称步冷曲线法)

热分析法一般用于具有较高相变温度的系统(如金属或合金系统)相图的测定。该方法是根据系统在加热或冷却过程中,温度随时间的变化情况来判断系统是否发生了相变。通常的做法是:先将样品全部熔化,然后让其在一定环境中自然冷却,并每隔一定时间(如 0.5 min 或 1 min)记录一次温度,以温度为纵坐标,时间为横坐标,作温度－时间曲线,此表示温度与时间关系的曲线叫作步冷曲线。当熔融系统在均匀冷却过程中无相变时,其温度将连续均匀下降,得到一平滑的冷却曲线;若在冷却过程中发生相变,则因系统产生的相变热与自然冷却时系统放出的热量相抵偿,温度降低变缓或停止,步冷曲线出现转折或水平线段,转折点所对应的温度,即为该组成的相变温度。利用冷却曲线所得到的一系列组成和所对应的相变温度数据,以横坐标表示混合物的组成,纵坐标表示温度,标出开始出现相变的温度,把这些点连接起来,就可绘出相图。

下面以二元简单低共熔系统相图的绘制为例进行介绍。图 2－2－3 为某一组成时系统的步冷曲线。当系统加热熔化后自然冷却时,由于无相变发生,体系的温度随时间变化较大,冷却较快,如图中 1—2 段所示。若冷却过程中系统某一组分发生相变,产生固相,放出凝固热,将使系统的冷却速度减慢,温度随时间的变化变得较平缓,如图中 2—3 段所示。当熔融液继续冷却到液相的组成为低共熔物的组成时,最低共熔混

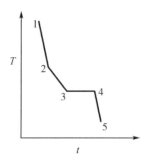

图 2－2－3 步冷曲线

合物完全凝固,此时系统温度保持不变,步冷曲线出现平台,如图中 3—4 段所示。当融熔液

完全凝固形成两种固体后,系统温度又继续下降,如图中 4—5 段所示。这样,根据一系列组成,做出一系列步冷曲线,然后以横坐标表示组成,以纵坐标表示温度,将组成与转折点、平台点相对应,在直角坐标系中描出一系列点,各转折点相互连接得某组分的溶解度曲线,各平台点相互连接得三相线,从而绘制出完整的相图,如图 2-2-4 所示。

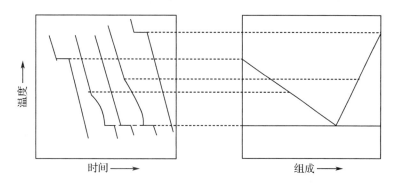

图 2-2-4　根据步冷曲线绘制相图

3. 溶解度法

溶解度法一般用于部分互溶系统相图的测定。下面以具有一对共轭溶液的三组分系统相图的绘制为例进行介绍。如图 2-2-5 所示,三组分相图中 A 和 B,A 和 C 为完全互溶,B 和 C 为部分互溶,曲线 bac 为溶解度曲线。曲线上方为单相区,曲线下方为二相区,物系点落在二相区内,即分为二相,如 X 点可分成组成为 E 和 F 的二相,而 EF 线称为连接线。

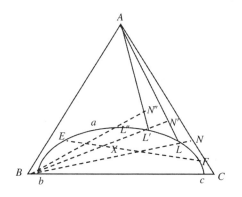

图 2-2-5　一对共轭溶液的三组分系统相图

（1）溶解度曲线的绘制:先选择完全互溶的两个组分(如 A 和 C),以一定的比例混合组成均相溶液,如图上的 N 点,滴加入组分 B,根据平衡相图的直线规则,物系点则沿着 NB 移动,直至溶液变浑,即为 L 点。再加入 A,物系点沿 LA 上升至 N' 点而变清。再加入 B,此时物系点又沿着 $N'B$ 由 N' 移动至 L' 而再次变浑,再加入 A……如此反复,最后连接 L、L'、L'' 等点即可画出溶解度曲线。

（2）连接线的绘制

由于连接线表示在两相区内呈平衡两相的组成(或 A 在两相中的分配),所以可以在两相区内配制溶液,待平衡后分析每相中的任何一种组成的含量,可通过连接溶解度曲线上该两含量的组成点而得出。

Experiment 2　Phase Diagram of Liquid-Vapor Equilibrium in a Binary System

【Purpose of the Experiment】

1. To draw a liquid-vapor equilibrium phase diagram in a binary system of ethanol and n-propanol.

2. To master the method of determining the binary liquid composition by refractive index.

【Principle】

1. Liquid-vapor phase diagram

A two-component system in which two liquid substances are mixed. If two components can soluble each other in any ratio, it is called a completely miscible two-liquid system. The boiling point of a liquid is the temperature at which the vapor pressure of the liquid is equal to the external pressure. At a certain external pressure, the boiling point of the pure liquid has a fixed value. However, the boiling point of the two-liquid system is not only related to the external pressure, but also related to the relative contents of the two liquids. According to the law: $F = C - P + 2$.

A vapor-liquid coexisting two-component system with $F = 2$. As long as a variable is arbitrarily determined, the existence state of the entire system can be described by a two-dimensional graph. For example, at a certain temperature, the relationship between the pressure of the system and the component can be plotted. On the $T - x$ phase diagram, there are three variables of temperature, liquid phase composition and gas phase composition, but only one degree of freedom. Figure 2 - 2 - 1 shows the n-propanol-ethanol system as an example. The horizontal line of temperature T_1 indicates the corresponding values of the liquid phase component x and the gas phase component y which are in equilibrium at this temperature.

Usually, the phase diagram of the liquid-vapor system can be drawn by measuring the boiling point of a series of different proportioning solutions and the composition of the vapor and liquid phases. Take Figure 2 - 2 - 1(a) as an example. Explain the principle of drawing the boiling point-composition phase diagram. Heating the solution with the total composition of x_1, the temperature of the system rises, and the solution begins to boil when it reaches the p point on the liquidus line. The vapor phase with the composition of y_1 starts to form, but the vapor phase is small (to 0), x_1, y_1 it represents the composition of liquid and vapor in two phases when equilibrium is reached. Continue heating, the vapor phase gradually increases, the boiling point continues to rise, the vapor and liquid two-phase composition changes in the vapor phase and the liquid phase respectively. When a certain temperature (such as q point) is maintained and the temperature is maintained, x_2, y_2 is at this temperature, the liquid and vapor phases are composed. The vapor and liquid two-phase samples were taken out separately, and the composition was analyzed to obtain the composition of each phase when the vapor and liquid phases were balanced at the temperature. The total composition of the solution was changed to obtain the composition of each phase when the vapor and liquid phases were balanced at another temperature. The vapor-liquid equilibrium temperature and the composition of vapor and liquid phase under the total composition of the solution were measured. The vapor phase points were connected by a vapor line, and the liquid phase points were connected by a liquid line to obtain a boiling point-composition phase diagram.

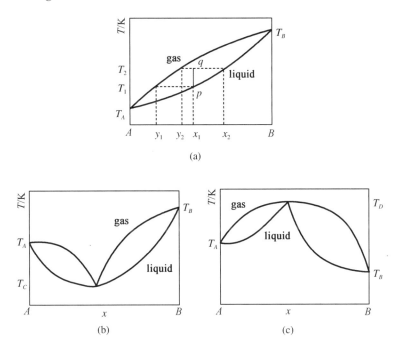

Figure 2-2-1 Boiling point-composition phase diagram of fully miscible two-liquid system

2. Ebulliometer

The boiling point instrument used in this experiment is shown in the Figure 2-2-2 ebulliometer.

3. Composition analysis

The composition of the solution was analyzed by a refractive index method. The principle of refractive index method: pure materials have a certain refractive index at a certain temperature. If the two liquids form a mixture, the refractive index varies with the composition. Thus, the refractive index (n) of several sets of known composition (x) solutions can be determined at a fixed temperature, and the $n-x$ isotherm (ie, the working curve) is plotted. For samples of unknown composition, after taking out each sample and measuring the refractive index (n) at that temperature (the temperature used to measure the standard solution n), the corresponding composition can be found from the working curve. Because the refractive indices of ethanol and n-propanol differ greatly, and the amount of sample required for refractive index method is small, it is suitable for this experiment. The commonly used instrument for measuring the refractive index is the Abbe refractometer.

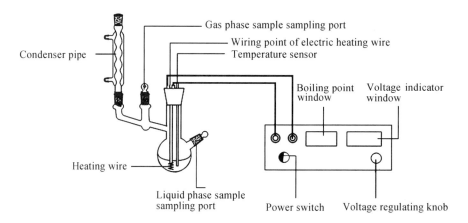

Figure 2-2-2 Ebulliometer

【Apparatus and Reagents】

Ebulliometer, 1; Abbe refractometer, 1; Pipet, 2; Sucker, 4; Droplet bottle, 18.
Ethanol (AR, Boiling point: 78.5 ℃); n-propanol (AR, Boiling point: 97.4 ℃).

【Procedures】

1. Working curve drawing

(1) Prepare various concentrations of ethanol-n-propanol standard solution according to Table 2-2-1 (note that the volume should be converted into weight when processing the data), put them into the numbered droplet bottle, and use Abbe to refraction. The refractive index of the instrument was measured, and the measurement was repeated three times and averaged.

(2) Draw the refractive index-compose the working curve at this temperature.

2. Sampling and measuring

(1) Open the "liquid sample sampling port" plug, add 12 mL ethanol, 8 mL propanol with pipet, then cover the glass plug, the condenser is connected to cold water, and turn on the boiling point power switch.

(2) Adjust the voltage to 15 V. After a period of time, the liquid begins to boiling, maintain the reflux state for about 5 min, and then read and record the boiling point of the liquid from the "boiling point indication window".

(3) Adjust the voltage to zero, let the system cool for 1 min, then take the condensate from the "vapor phase sample sampling port" with a sucker. After the extraction, the plug was stoppered, measure the refractive index of the sample with Abbe refractometer, and record.

(4) The liquid phase sample was taken from the "liquid sample sampling port" with a sucker. After the extraction, the plug was stoppered, and the refractive index was measured by Abbe refractometer and recorded.

(5) Open the "liquid sample sampling port" plug, add 8 mL of propanol with a pipet, cover with a glass stopper, and repeat (2) – (4) steps.

(6) Repeat step (5) four times.

【Records and Analysis of Data】

1. Fill the refractive index and composition of the ethanol-n-propanol standard solution in Table 2-2-1, and draw the working curve with the refractive index as the vertical axis and the composition as the horizontal axis.

2. Using the above working curve, determine the corresponding composition from the refractive indices of the vapor phase and liquid phase samples, and fill in Table 2-2-2.

3. Taking the boiling point as the vertical axis and plotting the composition of the vapor phase and the liquid phase as the horizontal axis, the boiling point-composition diagram of ethanol-n-propanol is plotted.

Table 2-2-1 Determination of refractive index of ethanol-n-propanol standard mixture

	number	1#	2#	3#	4#	5#	6#	7#	8#	9#	10#	11#
Mixture Liquid /mL	n-propanol	0	1	1	3	2	2	3	7	4	9	2
	ethanol	2	9	4	7	3	2	2	3	1	1	0
n-propanol $wt\%$												
n_D	1											
	2											
	3											
	average											

Table 2-2-2 Experimental data of ethanol-n-propanol binary liquid system

		number	1#	2#	3#	4#	5#	6#	7#
Liquid phase	Boiling point								
	Refractive index	1							
		2							
		3							
		average							
	Composition								

Table 2-2-2 (continue)

	number		1#	2#	3#	4#	5#	6#	7#
Vapor phase	Refractive index	1							
		2							
		3							
		average							
	Composition								

【Post Lab Questions】

1. What are the factors related to the refractive index of the solution?
2. What is the purpose of the refractive index-composition working curve of the ethanol-n-propanol standard solution?
3. How should the boiling point of pure ethanol and n-propanol be determined?

实验3 液体饱和蒸气压的测定

【实验目的】

1. 掌握用等压计测定在不同温度下纯液体(如无水乙醇)的饱和蒸气压;
2. 掌握用克劳修斯 – 克拉珀龙方程计算摩尔汽化焓;
3. 了解真空体系的设计、安装和操作的基本方法。

【预习要求】

1. 明确蒸气压、正常沸点、沸腾温度的含义;
2. 掌握用等压计测定流体饱和蒸气压的操作方法;
3. 了解真空泵、气压计的使用方法及注意事项;
4. 明确液体饱和蒸气压的测定原理及克劳修斯 – 克拉珀龙方程。

【实验原理】

在单组分物系发生变化时,饱和蒸气压与温度的关系,可用克拉珀龙(Clapeyron)方程来表示,即

$$\frac{\mathrm{d}p}{\mathrm{d}T} = \frac{\Delta_{\mathrm{vap}}H_{\mathrm{m}}}{T \cdot \Delta V} \tag{2-3-1}$$

克拉珀龙方程可应用于任何两相平衡。在应用于液 – 气两相平衡时,当温度变化不大时,摩尔汽化热 $\Delta_{\mathrm{vap}}H_{\mathrm{m}}$ 可近似看作常数。设蒸气为理想气体,并忽略液体的体积,可将上式积分得克劳修斯 – 克拉珀龙(Clausius – Clapeyron)方程,即

$$\lg p = \frac{-\Delta_{vap}H_m}{2.303R} \cdot \frac{1}{T} + C \tag{2-3-2}$$

式中,p 为液体在温度 $T(K)$ 时的蒸气压;C 为积分常数;R 为常量。

实验测得各温度下的饱和蒸气压后,以 $\lg p$ 对 $1/T$ 作图,得一直线,该直线的斜率 m 为

$$m = \frac{-\Delta_{vap}H_m}{2.303R} \tag{2-3-3}$$

由此即可求得摩尔汽化热 $\Delta_{vap}H_m$。

测定液体饱和蒸气压有三种方法。

(1)静态法:在某一温度下,直接测量饱和蒸气压。测量方法是调节外压与液体蒸气压相等,此法一般用于蒸气压比较大的液体。

(2)动态法:在不同外界压力下,测定液体的沸点。

(3)饱和气流法:使干燥的惰性气流通过被测物质,并使其为被测物质所饱和,然后测定所通过的气体中被测物质蒸气的含量,就可根据分压定律算出被测物质的饱和蒸气压。

本实验采用动态法利用等压计测定无水乙醇在不同外界压力下的沸点,即可得到该温度时的饱和蒸气压,实验装置如图 2-3-1 所示。通常一套真空体系装置由三部分构成:一是真空泵、缓冲储气罐部分,用以产生真空;二是真空的测量部分,包括数字压力计;三是实验部分,包括玻璃恒温水浴、U 形等位计、冷凝管。

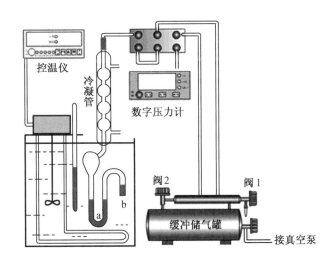

图 2-3-1 饱和蒸气压系统装置图

【仪器与试剂】

数字压力计 1 台,不锈钢缓冲储气罐 1 个,玻璃恒温水浴 1 台,真空泵 1 台,饱和蒸气压玻璃仪器(U 形等位计、冷凝管)1 套,橡胶管,电吹风 1 个,无水乙醇(分析纯),真空脂。

【实验步骤】

1. 按图 2-3-1 用橡胶管将各仪器连接成饱和蒸气压测定的实验装置。

2.检查体系装置气密性。

(1)缓冲储气罐整体气密性检查:将图2-3-2中的端口2处用橡胶管夹紧,再将进气阀及阀2打开,阀1关闭(三阀均为顺时针关闭,逆时针开启)。启动真空泵至压力计的数值为100~200 kPa,关闭进气阀,停止真空泵的工作。观察数字压力计的数值下降情况,若小于0.01 kPa/s,说明整体气密性良好,否则需查找并清除漏气原因,直至合格。

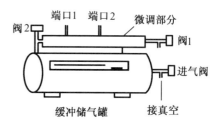

图2-3-2 缓冲储气罐示意图

(2)微调部分的气密性检查:关闭真空泵、进气阀及阀2,用阀1调整微调部分的压力,使其低于储气罐中压力的1/2,观察数字压力计,其变化值在标准范围内(小于0.01 kPa/4 s),说明气密性良好。若压力值上升超过标准,说明阀2泄漏;若压力值下降超过标准,说明阀1泄漏。

(3)被测系统的气密性检查:松开端口2处橡胶管,使之与被测系统连接。关闭阀1,开启阀2,使微调部分与罐内压力相等。之后,关闭阀2,开启阀1,观察数字压力计,显示值变化小于等于0.01 kPa/4 s即为合格。检漏完毕,开启阀1使微调部分泄压至零。

3.装样品。拔下等压计磨口连接管,从等压计加料口装入纯液体(如无水乙醇),使之充满图2-3-1中U形等压计a的2/3和b的大部分,然后连接磨口连接管,并安装入恒温水浴中,固定。连接磨口连接管时要注意密封性。

4.测定过程如下:

(1)打开恒温水浴开关,调节恒温水浴水温为T_1℃,待温度恒定10 min后开始实验。

(2)打开阀1,在与大气连通的情况下设置数字压力计,将数字压力计单位调为"kPa",按置零键置零。

(3)关闭阀1,打开阀2与进气阀,打开真空泵,观察数字压力计读数,等读数为-90 kPa以下时,关闭阀2和进气阀活塞,关闭真空泵。

(4)慢慢旋转与空气相连的活塞(阀1),缓慢放空气进入体系,使U形等压计液面水平,若2 min内保持液面水平不变,记下数字压力计的读数。

(5)关闭阀1,调节恒温水浴温度为T_1+5(℃),并恒温10 min。调节阀1与阀2使U形等压计液面水平,若2 min内保持液面水平不变,记下T_1+5(℃)时数字压力计的读数。共测十组不同温度下的饱和蒸气压数据。

(6)待所有数据测完后,打开阀1和阀2,打开进气阀,关闭恒温水浴及真空测压仪电源,实验结束。

【实验记录及数据处理】

1.将温度、压力数据列于表2-3-1中,算出不同温度的饱和蒸气压。

2.作$\lg p - 1/T$图,求出此直线的斜率,由斜率算出被测液体在实验温度范围内的平均摩尔汽化热$\Delta_{vap}H_m$。

被测液体：_____   室温：_____ ℃   大气压：_____ kPa

表 2-3-1　饱和蒸气压实验数据

温度计读数		$1/T \times 10^3$ /K^{-1}	压差计读数 $\Delta p/\text{kPa}$	饱和蒸气压 $p = p_{大气} - \Delta p$	$\lg p$
$t/℃$	T/K				

【讨论与说明】

1. 旋转与空气相连的活塞和调平 U 形等压计液面应缓慢进行，否则一旦进入过多空气需重新抽真空。
2. 标准口涂真空油脂时，只能涂下面 2/3 左右，以免沾污被测系统。
3. 开、关真空泵前，均应将其与大气接通。
4. 数字压力计使用前需归零。

【思考题】

1. 用动态法测定液体饱和蒸气压有何优缺点？
2. 说明饱和蒸气压、正常沸点和沸腾温度的含义。
3. 本实验主要系统误差有哪些？
4. 克拉珀龙方程在哪些情况下才能应用？
5. 缓冲瓶在实验中起什么作用？
6. 汽化热与温度有无关系？
7. 能否在加热情况下检查是否漏气？

【相关阅读】

真 空 技 术

真空是泛指低于标准压力的气体状态。在真空下，由于气体稀薄，单位体积内的分子数较少，分子间碰撞或分子在一定时间内碰撞于器壁的次数也相应减少，这就是真空的主要特点。

真空度是对气体稀薄程度的一种客观量度，其最直接的物理量应该是单位体积中的分子数。不同真空状态体现该空间具有不同的分子密度。但由于历史沿革，真空度的高低通常用气体的压强来表示，气体的压强越低表示真空度越高。

早期，人们曾将 760 mmHg 规定为标准压力，后来发现汞有七种同位素，同一汞柱表现的压强并不是唯一的，要视其同位素的比例而定，所以第十届国际计量大会决定以帕斯卡(Pascal)来定义标准压力，即规定 101.325 kPa 为标准压力。现行的 SI 中，真空度的单位和压强的单位均统一为帕，符号 Pa。

真空区域的划分，国际上尚未有统一的规定。在物理化学实验中通常按真空的获得和测量方法的不同，将真空划分为以下五个区域。

①粗真空：$10^2 \sim 10^3$ Pa，分子相互碰撞为主，分子自由程 $\lambda \ll$ 容器尺寸 d。
②低真空：$10^2 \sim 10^{-1}$ Pa，分子相互碰撞和分子与器壁碰撞不相上下，$\lambda \approx d$。
③高真空：$10^{-1} \sim 10^{-6}$ Pa，分子与器壁碰撞为主，$\lambda \gg d$。
④超高真空：$10^{-6} \sim 10^{-10}$ Pa，分子与器壁碰撞次数亦减小，形成一个单分子层的时间已达数分钟或数小时。
⑤极高真空：$<10^{-10}$ Pa，分子数目极为稀少，以致统计涨落现象较严重，与经典的统计理论产生偏离。

在近代的物理化学实验中，凡是涉及气体的物理化学性质、气体反应动力学、气固吸附以及表面化学研究，为了排除空气和其他气体的干扰，通常都需要在一个密闭的容器内进行，并且首先将干扰气体抽去，创造一个具有某种真空度的实验环境，然后将被研究的气体通入，才能进行有关研究。因此，真空的获得和测量是物理化学实验技术的一个重要方面。

用来产生真空的设备统称为真空泵。由于真空的区域为 $10^4 \sim 10^{-13}$ Pa，达十几个数量级的宽广范围，所以产生不同真空度时常采用不同种类的真空泵。高真空或超高真空的获得一般需用几种泵的组合系统。

真空泵的种类繁多，但按抽气的机理和方式可分为两大类。一类是压缩型的真空泵，将气体由特定容器压缩，然后排放至容器外。如利用高速水流将气体带走的水冲泵，利用膨胀-压缩作用的机械泵和隔膜泵，利用气体黏滞牵引作用的蒸气流喷射的扩散泵，以及利用分子牵引作用的分子泵等都属于压缩型真空泵。另一类是吸附型真空泵，靠吸附方式降低特定空间的气体分子密度。如利用活性表面吸附气体的钛泵，利用深冷表面使气体碰撞黏附的低温泵，以及利用吸附剂降低气体分子密度的吸附泵等均为吸附型真空泵。

机械泵和扩散泵都要用特种油作为工作物质，因而对实验对象有一定的污染，但由于这两种泵价格较低，它们在实验室中还经常被使用。机械泵的抽气速率很高，但只能产生 $1 \sim 0.1$ Pa 的低真空。扩散泵使用时必须用机械泵作为前级泵，可获得 $10^{-1} \sim 10^{-7}$ Pa 的高真空或超高真空。隔膜泵、吸附泵和钛泵都属于无油类型泵，不存在油蒸气的玷污问题，它们串级使用可获得优于 10^{-6} Pa 的超高真空。分子泵是靠内圆筒高速机械运动使气体做定向流动，一般须与前级泵组合来获得高真空。低温泵是目前抽速最大，能达到极高真空的真空泵。

1. 水冲泵

水冲泵的结构如图 2-3-3 所示。水经过收缩的喷口以高速喷出，使喷口处形成低压，产生抽吸作用将由系统进入的气体分子不断被高速喷出的水流带走。水冲泵能达到的极限真空受水的蒸气压所限制,20 ℃时极限真空约为 10^3 Pa。水冲泵在实验室主要用于产生粗真空。

2. 机械泵

常用的机械泵为旋片式泵,一般由两个抽气级前后串联构成。每一级均由泵体、转子和滑片组成。转子在旋转时始终紧贴泵体缸壁上。镶在转子槽中的滑片靠弹簧的压力也紧贴缸壁,由此使泵体的进、排气部分被转子和滑片分隔成两个部分:进气部分和排气部分。当转子旋转时,进、排气部分随着滑片的伸缩,它们的体积周期性地扩大和压缩。进气部分因体积扩大而压力降低,起着吸气作用。排气部分因体积压缩,起着排气作用。由于第一级的排气口与第二级的进气口串

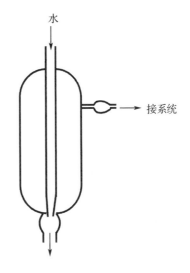

图 2 - 3 - 3　水冲泵的结构示意图

联,使第一级排出的气体由第二级吸入,再由第二级的排气塞排出泵外。这种串联结构的优点是使直接与被抽容器相接的第一级的反压力得到显著的降低,使泵能达到较大的真空度。

为了减少漏气和润滑冷却机件,其排气阀以下部分全部浸在蒸气压较低且又有一定黏度的机械泵油内。

使用机械泵必须注意:①如被抽气的实验系统有易凝结的蒸气或挥发性液体或腐蚀性气体,则不能直接用机械泵对系统抽气,应在系统和泵的进气管之间串联冷凝器、洗气瓶或吸收瓶,以除去上述气体;②泵的进气管前要连接一个三通活塞或加装与机械泵同步开关的电磁阀。在机械泵停止运行前,应先通过三通活塞使泵的进气口与大气相通或通过电磁阀散气使两边压力平衡,以防泵油倒吸污染实验系统。

3. 扩散泵

扩散泵的原理是利用一种工作物质高速从喷口处喷出,在喷口处形成低压,对周围气体产生抽吸作用而将气体带走。这种工作物质在常温时应是液体,并具有极低的蒸气压,用小功率的电炉加热就能使液体沸腾汽化,沸点不能过高,通过水冷却便能使汽化的蒸气冷凝下来。过去用汞作为工作物质,但因汞有毒,现在通常采用硅油。硅油被电炉加热沸腾汽化后,通过中心导管从顶部的二级喷口处喷出,在喷口处形成低压,将周围气体带走,而硅油蒸气随即被冷凝成液体回入底部,循环使用。被夹带在硅油蒸气中的气体在底部聚集,立即被机械泵抽走。

在上述过程中,硅油蒸气起着一种抽动作用,其抽运气体的能力决定下面三个因素:硅油本身的物质的量要大,喷射速度要高,喷口级数要多。现在用物质的量大于 3 000 以上的硅油做工作物质的四级扩散泵,其极限真空度可达 10^{-7} Pa,三级扩散泵可达 10^{-4} Pa。实验室用的油扩散泵的抽气速率通常有 60×10^{-3} $m^3 \cdot s^{-1}$ 和 300×10^{-3} $m^3 \cdot s^{-1}$ 两种。

油扩散泵必须用机械泵作为前级泵,将其抽出的气体抽走,不能单独使用。扩散泵的硅油易被空气氧化,所以使用时应用机械泵先将系统抽至低真空后,才能加热硅油。硅油不能承受高温,否则会裂解。硅油蒸气压虽然极低,但仍会蒸发一定数量的油分子进入真空系

统,污染被研究对象,因此一般在扩散泵和真空系统连接处安装冷凝阱,以捕捉可能进入系统的油蒸气。

4. 分子泵

分子泵是一种纯机械的、高速旋转的真空泵,高速旋转($10\,000 \sim 75\,000\ \mathrm{r \cdot min^{-1}}$)的涡轮叶片,不断对被抽气体施以定向的动量和牵引压缩作用,将气体排走。分子泵的动轮叶与静轮叶间距仅数毫米,两者相间排列,且叶面角相反,从而达到最大的抽气作用。由于轴承上有润滑油,故分子泵不是严格的无油泵。目前已制成一种磁悬浮轴承,转速更高,可方便地获得小于 $10^{-8}\ \mathrm{Pa}$ 的超高真空。

5. 吸附泵

吸附泵的全名为分子筛吸附泵。它是利用分子筛在低温时能吸附大量气体或蒸气的原理制成的,其特点是将气体捕集在分子筛内,而不是将气体排出泵外。分子筛是人工合成的无水硅铝酸盐结晶,其内部充满着孔径均匀的无数微孔空穴,约占整个分子筛的一半。当向液氮筒中灌入液氮后,分子筛因被冷却到低温,能大量捕集待抽容器中的气体,极限真空度可达约 $10^{-1}\ \mathrm{Pa}$。由于吸附后的分子筛可通过加热脱附活化,可反复使用,因此吸附泵的使用寿命较长,维护方便。吸附泵可单独使用,其优点是无油,但工作时需消耗液氮。通常吸附泵用作超高真空系统中钛泵的前级泵。

6. 钛泵

钛泵的抽气机理通常被认为是化学吸附和物理吸附的综合,且以化学吸附为主。钛泵的种类很多,这里介绍一种磁控管型冷阴极溅射离子钛泵。其阳极为不锈钢圆筒,阴极为钛柱,两极之间加 $3 \sim 6\ \mathrm{kV}$ 高电压。泵壳由不锈钢制成,置于 $0.13 \sim 0.15\ \mathrm{T}$ 的磁场中。泵内空间的游离电子在电场作用下被阳极加速,将遇到的气体分子电离,形成的正离子打在阴极钛柱上,产生二次电子。这些从冷阴极发射出来的二次电子,径向被拉向阳极,同时受磁场作用发生偏转,因此沿轮滚线轨迹绕轴运动,并不断撞击所遇到的气体分子,使气体电离,形成潘宁放电。放电后,圆筒内气体电离所产生的正离子因质量大而不受磁场约束,在电场作用下迅速撞击阴极,在阴极上引起钛溅射。溅射的活性钛在圆筒内各处形成钛膜,大量吸附泵壳内的气体分子。气体电离所产生的电子同样按轮滚线轨迹绕轴运动,使所撞击的气体分子电离。

还有一种钛升华泵,原理是利用大电流引起的高温使钛蒸发,利用活性的钛原子与气体分子反应或包埋惰性气体分子来达到减少空间内气体分子数目,即提高真空的作用。

钛泵不能单独使用,需用吸附泵或机械泵作为其前级泵。钛泵具有极限真空度高(约 $10^{-8}\ \mathrm{Pa}$)、无油、无噪声和无振动等优点,在 $10^{-2}\ \mathrm{Pa}$ 时仍有较高的抽速,且操作简便,使用寿命长。

7. 低温泵

低温泵是目前抽速最高,能达到真空的泵。其原理是靠深冷表面抽气,主要用于产生模拟的宇宙环境。这种泵的抽气速率正比于深冷表面的大小,一般可达 $10^3\ \mathrm{m^3 \cdot s^{-1}}$。

在液氦温度(4.2 K)下,许多气体的蒸气压几乎为零,因此低温泵可获得 $10^{-9} \sim 10^{-10}\ \mathrm{Pa}$ 的超高真空或极高真空。但氢气是个例外,液氦温度下其蒸气压为 $10^{-4}\ \mathrm{Pa}$。如果在低温泵的进口处加一催化泵使氢电离,再经氧化铜使氢变成水,就可以解决抽氢气的困难。在液氦板槽外加液氮的板槽,是为了防止热辐射。

Experiment 3 Determination of Saturated Vapor Pressure of a Pure Liquid

【Purpose of the Experiment】

1. Determine the saturated vapor pressure of a pure liquid (anhydrous ethanol) at different temperatures using an isobaric meter.
2. Master the calculation of molar vaporization by Clausius-Clapeyron equation.
3. Understand the basic methods of designing, installing and operating a vacuum system.

【Preview Requirement】

1. Defining the meaning of vapor pressure, normal boiling point and boiling temperature.
2. Master the method of measuring the saturated vapor pressure of a liquid with an isobaric meter.
3. Understand the use of vacuum pumps and barometers and precautions.
4. The principle of determination of liquid saturated vapor pressure and the Clausius-Clapeyron equation are clarified.

【Principle】

When the one-component system changes, the relationship between saturated vapor pressure and temperature can be expressed by the Clapeyron equation.

$$\frac{dp}{dT} = \frac{\Delta_{vap} H_m}{T \Delta V} \quad (2-3-1)$$

The Clapeyron equation can be applied to any two-phase equilibrium. When it is applied to the liquid-gas two-phase equilibrium, the molar enthalpy of vaporization ($\Delta_{vap} H_m$) can be approximated as a constant when the temperature does not change much. If we suppose the vapor is the ideal gas and ignore the volume of the liquid, the upper formula can be integrated into the Clausius-Clapeyron equation.

$$\lg p = \frac{-\Delta_{vap} H_m}{2.303 R} \cdot \frac{1}{T} + C \quad (2-3-2)$$

Where p is the vapor pressure of the liquid at temperature $T(K)$; C is the integral constant; R is the constant.

After measuring the saturated vapor pressure at each temperature, the graph is plotted as $\lg p$ versus $1/T$, and the straight line is obtained. The slope m of the straight line is:

$$m = \frac{-\Delta_{vap} H_m}{2.303 R} \quad (2-3-3)$$

From the above equation, the molar heat of vaporization can be obtained.

There are three ways to determine the liquid saturated vapor pressure.

(1) Static method: directly measure the saturated vapor pressure at a certain temperature. The measurement method is that the external pressure is equal to the liquid vapor pressure, and the method is generally used for a liquid having a relatively large vapor pressure;

(2) Dynamic method: determine the boiling point of the liquid under different external pressures;

(3) Saturated airflow method: the dry inert gas stream is passed through the tested substance and saturated with the substance, and then the vapor content of the tested substance in the passed gas is measured, and the saturated vapor pressure of tested substance can be calculated by law of partial pressure.

In this experiment, the boiling point of anhydrous ethanol under different external pressures is determined by dynamic method using an isobaric meter, and the saturated vapor pressure at this temperature is obtained. The experimental apparatus is shown in Figure 2-3-1. A set of system consists of three parts: one is a vacuum pump, a buffer tank part to generate vacuum; the other is the vacuum measurement part, including the digital low vacuum pressure gauge; the third is the experimental part, including glass thermostatic water bath, U-shaped equal gauge, condensing tube.

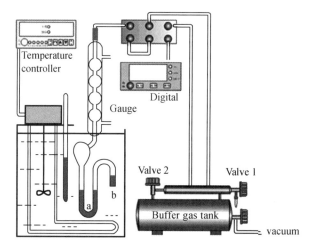

Figure 2-3-1 The device diagram of saturated vapor pressure system

【Apparatus and Reagents】

Digital pressure gauge; Stainless steel buffer gas tank; Glass constant temperature water bath; Vacuum pump; Saturated vapor pressure glass instrument; Rubber hose; Electric drier; Absolute ethyl alcohol; Vacuum grease.

【Procedures】

1. The instruments were connected with rubber tubing as an experimental device for the measurement of saturated vapor pressure, as shown in Figure 2-3-1.

2. Check the air tightness of the system.

(1) Buffer gas tank overall air tightness inspection: Clamp the rubber tube at port 2 in Figure 2-3-2, then open the inlet valve and valve 2, and close the valve 1 (The three valves are closed clockwise and counterclockwise). Start the vacuum pump to the pressure gauge number from 100 kPa to 200 kPa, close the inlet valve and stop the vacuum pump. Observe the drop in the value of the digital manometer. If it is less than 0.01 kPa/s, the overall air tightness is good. Otherwise, find and clear the cause of the leak until it is qualified.

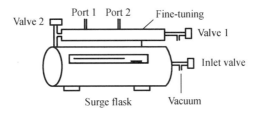

Figure 2-3-2 Schematic diagram of buffer gas storage tank

(2) Air tightness check for fine tuning part: Close vacuum pump, inlet valve and valve 2. Use valve 1 to adjust the pressure in the fine-tuning part, in order to let it lower than 1/2 of the pressure in the gas storage tank. Observe the digital manometer, when the change value is within the standard range (less than 0.01 kPa/4 s), indicating good air tightness. If the pressure value rises above the standard, it indicates valve 2 leakage. If pressure drops more than standard, valve 1 leaks.

(3) Air tightness check of the system under test: Loosen the rubber hose at port 2 to connect it to the system. Close valve 1, open valve 2, and make the fine-tuning part equal to the pressure in the tank. After that, shutoff valve 2 and open valve 1. Observe the digital manometer, when the value changes ≤0.01 kPa/4 s, which is qualified. After leak detection, open valve 1 to reduce pressure to zero.

3. Loading sample. Remove the grinding pipe of the isobaric meter and fill it with pure liquid (such as anhydrous ethanol) from the feed port of the isobaric meter. Fill it with 2/3 of the U type isobaric meter a and most of the U type isobaric meter b. Then, connect the original grinding mouth connection pipe, and install into the constant temperature water bath. Pay attention to the tightness when connecting the grinding joint.

4. Determination

(1) Open the constant temperature water bath switch, adjust the water temperature of the constant temperature bath to T_1 (℃), and start the experiment after the temperature is constant for 10 min.

(2) Open valve 1, set the digital pressure gauge in communication with the atmosphere, adjust the digital pressure gauge unit to kPa, and press the zero button to zero.

(3) Shutoff valve 1, open valve 2 and inlet valve, open vacuum pump, observe digital manometer reading, when the reading is below −90 kPa, shutoff valve 2 and inlet valve piston, close vacuum pump.

(4) Slowly rotate the piston (valve 1) connected to the air and slowly put the air into the system to make the level of the U type isobaric meter level. If the level remains unchanged within 2 min, write down the reading of the digital manometer.

(5) Shutoff valve 1. Adjust the temperature of constant temperature water bath to $T_1 + (5\ ℃)$ and keep it constant for 10 min. Control valve 1 and valve 2 to keep the level of U type isobaric meter. If the level remains unchanged for 2 min, record the reading of digital manometer at $T_1 + 5(℃)$. Ten sets of saturated vapor pressure data at different temperatures were measured.

(6) After all the data are measured, open valve 1 and valve 2, open the air inlet valve, close the constant temperature tank and vacuum pressure measuring instrument source, and the experiment is completed.

【Records and Analysis of Data】

1. Figure out the saturated vapor pressure at different temperatures by tabulated temperature and pressure data in Table 2−3−1.

2. Make $\lg p - 1/T$ diagram, find the slope of this line, and calculate the $\Delta_{vap}H_m$ of the measured liquid in the experimental temperature range from the slope.

Tested liquid:_____   atmospheric pressure:_____

Table 2−3−1 Experimental data of saturated vapor pressure

Temperature		$1/T \times 10^3$ /K^{-1}	Δp/kPa	$p = p_{DQ} - \Delta p$	$\lg p$
$t/℃$	T/K				

【Notes】

1. The U type isobaric meter level should be leveled slowly by rotating the piston connected to air, otherwise it needs to be vacuumized again once too much air enters.

2. When applying vacuum grease to the standard mouth, only about 2/3 of the bottom should be applied to avoid contamination of the tested system.

3. Before turning on and off the vacuum pump, it should be connected to the atmosphere.

4. The digital manometer should be reset to zero before use.

【Post Lab Questions】

1. What are the advantages and disadvantages of dynamic measurement of liquid vapor pressure?

2. Explain the meaning of saturated vapor pressure, normal boiling point and boiling temperature.

3. What are the main systematic errors in this experiment?

4. Do you know under what circumstances the clapeyron equation can be applied?

5. What does the buffer bottle do here?

6. Is the $\Delta_{vap}H_m$ related to temperature?

7. Is it possible to check for air leakage under heating conditions?

实验 4　氨基甲酸铵分解反应平衡常数的测定

【实验目的】

1. 熟悉用等压法测定氨基甲酸铵的分解压力，计算分解反应平衡常数及有关热力学函数；

2. 掌握真空实验技术。

【预习要求】

1. 理解平衡常数的定义；

2. 了解平衡常数的测定方法。

【实验原理】

氨基甲酸铵是合成尿素的中间产物，为白色固体，很不稳定，易分解。在一定温度下它的分解平衡反应式为

$$NH_2COONH_4(s) \rightleftharpoons 2NH_3(g) + CO_2(g)$$

在实验条件下可将气体按照理想气体处理，在常压下其平衡常数 $K^\ominus$ 可近似表示为

$$K^{\ominus} = \left[\frac{p_{NH_3}}{p^{\ominus}}\right]^2 \left[\frac{p_{CO_2}}{p^{\ominus}}\right] \qquad (2-4-1)$$

其中,p_{NH_3} 和 p_{CO_2} 分别表示反应温度下氨气和二氧化碳平衡时的分压;体系总压 $p = p_{NH_3} + p_{CO_2}$。由化学反应计量方程式可知

$$p_{NH_3} = \frac{2}{3}p, \quad p_{CO_2} = \frac{1}{3}p \qquad (2-4-2)$$

将式(2-4-2)代入式(2-4-1)得

$$K^{\ominus} = \left(\frac{2p}{3p^{\ominus}}\right)^2 \left(\frac{p}{3p^{\ominus}}\right) = \frac{4}{27}\left(\frac{p}{p^{\ominus}}\right)^3 \qquad (2-4-3)$$

因此,当体系平衡后,测量其总压 p,即可计算出平衡常数 $K^{\ominus}$。

温度对平衡常数的影响可用下式表示,即

$$\ln K^{\ominus} = -\frac{\Delta_r H_m^{\ominus}}{RT} + C' \quad (C' \text{ 为积分常数}) \qquad (2-4-4)$$

若以 $\ln K^{\ominus}$ 对 $1/T$ 作图,得一直线,其斜率为 $-\Delta_r H_m^{\ominus}/R$,由此可求出 $\Delta_r H_m^{\ominus}$,可按下式计算 T 温度下反应的标准吉布斯自由能变化 $\Delta_r G_m^{\ominus}$

$$\Delta_r G_m^{\ominus} = -RT\ln K_p^{\ominus} \qquad (2-4-5)$$

利用实验温度范围内反应的平均等压热效应 $\Delta_r H_m^{\ominus}$ 和 T 温度下的标准吉布斯自由能变化 $\Delta_r G_m^{\ominus}$,可近似计算出该温度下的熵变 $\Delta_r S_m^{\ominus}$,即

$$\Delta_r S_m^{\ominus} = \frac{\Delta_r H_m^{\ominus} - \Delta_r G_m^{\ominus}}{T} \qquad (2-4-6)$$

因此,通过测定一定温度范围内某温度的氨基甲酸铵的分解压(平衡总压),就可以利用上述公式分别求出 $K^{\ominus}$、$\Delta_r H_m^{\ominus}$、$\Delta_r G_m^{\ominus}$、$\Delta_r S_m^{\ominus}$。

【仪器与药品】

仪器:分解压力测定仪(等压计、缓冲压力罐)1 套,恒温水浴 1 套,低真空数字测压仪 1 台,真空泵 1 台。

试剂:氨基甲酸铵(分析纯),液体石蜡。

【实验步骤】

1. 读取大气压和室温

从气压计读出大气压 $p_{大气}$,实验前后各读一次取平均值。

2. 装样

如图 2-4-1 所示,将氨基甲酸铵粉末从图示所在位置装入盛样的小球(加入固体约 0.5 g),然后从图示加液体石蜡处向 U 形管中加入液体石蜡作液封(液体石蜡填充 U 形管约一半的高度),最后将等压管与真空系统连接并密封好。

3. 压力计"采零"

按图 2-4-1 所示接好测量仪器,开启低真空数字压力计,电源预热 2 min,压力计的压力显示单位切换到"kPa",打开进气阀,待压力计示数稳定后,按"采零"键,以消除仪表系统的零点漂移,此时 LED 显示为"0000"。

4.检查装置气密性

(1)缓冲储气罐整体气密性检查:将图2-4-1中的橡胶管夹拧紧,然后关闭进气阀,打开平衡阀和抽气阀(三阀均为顺时针关闭,逆时针开启)。启动真空泵至压力计数字为100~200 kPa,关闭抽气阀,停止真空泵的工作,观察数字压力计的数值下降情况,若压力计读数增速小于0.01 kPa/s,说明整体气密性良好。否则需查找并清除漏气原因,直至气密性良好。

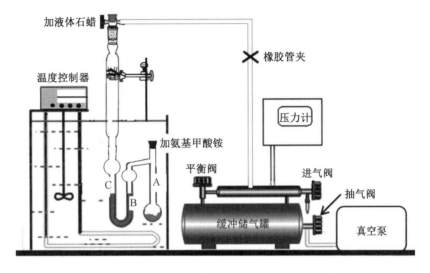

图2-4-1 氨基甲酸铵分解平衡测量装置图

如判断此时漏气,关闭平衡阀,压力计读数增速大于0.02 kPa/s,则进气阀漏气的可能性大,反之则是抽气阀或平衡阀漏气。

(2)被测系统的气密性检查:确定缓冲储气罐不漏气后,打开橡胶管夹开关,使被测系统与缓冲储气罐相连。观察压力计读数增速小于0.01 kPa/s,即表明气密性良好;反之,检查橡胶管及各接口处,直至气密性合格为止。检漏完毕,开启进气阀使微调部分泄压至零。

5.分解压力的测定

调节恒温水浴温度至35 ℃,当水浴温度恒定至35 ℃,关闭进气阀,打开抽气阀和平衡阀,启动真空泵对系统进行抽气,压力计读数从0减至约-90 kPa时,关闭进气阀,随后关闭真空泵。这时U形等压管中两管液面不处于同一水平面,C管液面明显高于B管液面。缓慢打开进气阀增大系统压力,切忌快速打开进气阀,以免B管中液体倒灌入A管中,持续调节进气阀使B管液面和C管液面保持同一水平,且能恒定2 min以上,表明A管中氨基甲酸铵达到分解平衡,记录此时压力计的读数。需注意的是:压力计的读数并不是氨基甲酸铵分解压,需要进一步用此读数计算后才能获得氨基甲酸铵的分解压。

再次打开抽气泵减小系统的压力,重复步骤5中的上述操作,测平行数据,要求两次测量的压力差小于0.3 kPa,记录压力计显示的压力,求平均值。

按照步骤5继续测定40 ℃、45 ℃、50 ℃的分解压力。

6.实验结束

缓慢打开进气阀(防止B管中液体倒灌入A管),放入空气,使被测系统泄压至零(缓慢关闭平衡阀,打开进气阀),取下等压管,关闭所有电源,整理好实验台面。

【数据记录及处理】

1. 将实验数据记入表 2-4-1。

室温:_____℃   大气压:_____kPa

表 2-4-1 分解压及平衡常数实验数据

温度			压力计读数	系统平衡压力	平衡常数	
$t/℃$	T/K	$1/T \times 10^3/K^{-1}$	$\Delta p/kPa$	$p = 大气压 - \|\Delta p\|$	$K^{\ominus}$	$\ln K_p^{\ominus}$

2. 以 $\ln K^{\ominus}$ 对 $1/T$ 作图,由所得直线的斜率根据公式求 $\Delta_r H_m^{\ominus}$。

3. 按照有关的热力学公式,计算 25 ℃时氨基甲酸铵的 $\Delta_r G_m^{\ominus}$ 和 $\Delta_r S_m^{\ominus}$。实验中所需要热力学数据参照表 2-4-2。

4. 将实验测得的氨基甲酸铵分解压与表 2-4-3 中数据做比较,计算其相对误差。

【实验注意事项】

1. 真空泵停泵前先使泵的进气阀与大气相通,以防油倒吸污染实验系统。

2. 实验过程中打开进气阀和平衡阀时一定要缓慢进行,关闭进气阀和平衡阀时一定要关紧,这是保证实验顺利进行的重要操作之一。

3. 实验测定后几个温度的分解压时,不需要再重新抽气,只需要调节液面相平。

4. 本实验的装置与测定液体饱和蒸气压的装置相似,所以本装置可用来测定液体的饱和蒸气压。

【思考题】

1. 什么叫分解压?在一定温度下,氨基甲酸铵的用量多少对分解压力有何影响?
2. 怎样测定氨基甲酸铵的分解压力,样品的数量对分解平衡总压有无影响?
3. 如何判断氨基甲酸铵分解已达平衡,如没有平衡会对测出的数据有何影响?
4. 当使空气通入系统时,若通得过多有何现象出现,如何克服?

【文献参考值】

文献参考值见表 2-4-2 和表 2-4-3。

表 2-4-2　实验文献热力学数据参考值

项目	$NH_2COONH_4(s)$	$NH_3(g)$	$CO_2(g)$
$\Delta_f H_m^\ominus/(kJ\cdot mol^{-1})$	-645.51	-46.141	-393.792
$\Delta_f G_m^\ominus/(kJ\cdot mol^{-1})$	-448.386	-16.497	-394.642
$S_m^\ominus/(J\cdot K^{-1}\cdot mol^{-1})$	133.565	192.476	213.788

表 2-4-3　不同温度下氨基甲酸铵的分解压

温度/℃	25	30	35	40	45	50
$p_{分解}$/kPa	11.73	17.07	23.80	32.93	45.33	62.93

【相关阅读】

1. 平衡常数的测定

平衡常数可以用实验方法测定,也可以利用热力学数据计算而得。

实验方法通常有物理方法和化学方法。

(1) 物理方法

通过测定平衡系统的物理性质,然后导出平衡常数,如测定系统的折光率、电导率、压力、旋光度或体积的改变等。这种方法要求系统的被测物理性质须与有关反应物质的浓度变化成单调函数关系,且达到平衡时所测的物理性质不再发生变化。物理方法的优点是在测定时不会干扰或破坏体系的平衡状态,也不需停止反应,可进行在线检测。

(2) 化学方法

利用化学分析的方法直接测定平衡系统有关物质的浓度时,为了能较准确地测得平衡时的浓度,防止外界因素的干扰,在分析前就必须使平衡"冻结"。通常采取骤然降低温度的方法,使平衡"冻结",然后在低温下进行分析。若反应是催化反应,则可以移去催化剂使反应"停止",再进行测定。对液相反应还可以加入大量的溶剂把溶液冲稀,以降低平衡移动速率。例如,要测定反应 $2H_2+O_2 \Longrightarrow 2H_2O$ 在 2 000 ℃ 达到平衡时的平衡常数,可以将一定量的水置于耐高温的合金管中加热,在 2 000 ℃ 时保持一段时间,使之达到化学平衡。然后,将管子骤然冷却,再分析其中 H_2O、H_2、O_2 的含量,便可计算出在 2 000 ℃ 时这个反应的平衡常数。

(3) 如何判断反应已达平衡状态

根据化学平衡的特征,如果反应条件不改变,当反应达平衡态时,不论取样时间间隔多长,被测物质的浓度不变。或者保持其他条件不变,任意改变参加反应物质的初始浓度,若测得的平衡常数相同,则反应已达平衡。也可以先从正向反应开始测定,再从逆向反应测定,如测得的平衡常数相等,就证明反应已达平衡状态。

我们以合成氨的发展史为例来体现研究化学平衡的重要意义。在合成氨的发展史中除了化学平衡的重要性之外,化学动力学的重要意义也在其发展史中得以体现。

2. 合 成 氨

氨是一种基本化工原料,随着农业发展和军工生产的需要,20世纪初先后开发并实现了氨的工业生产。从氰化法演变到合成氨法,可以说合成氨的历史清晰地反映了化学平衡和化学动力学的重要性。

(1) 早期氰化法

1898年,德国的弗兰克等人发现空气中的氮能被碳化钙固定而生成氰氨化钙(又称石灰氮),进一步与过热水蒸气反应即可获得氨,即

$$CaCN_2 + 3H_2O \longrightarrow 2NH_3 + CaCO_3$$

1905年,德国氮肥公司建成世界上第一座生产氰氨化钙的工厂,这种制氨方法被称为氰化法。第一次世界大战期间,德国、美国主要采用该法生产氨,满足了军工生产的需要。氰化法固定每吨氮的总能耗为153 GJ,由于成本过高,到20世纪30年代被淘汰。

(2) 合成氨法

随着农业的发展和军工生产的需要,迫切要求建立转化率高、产量大、成本低的新的氨合成方法。于是研究人员设想,把空气中大量的氮气固定下来,以氮和氢为原料合成氨。最先研究氢气和氮气在高压下直接合成氨的反应的是法国化学家勒沙特利(Henri Le ChateLier,1850—1936)。勒沙特利于1900年做了相关实验,可惜的是他所用的氢气和氮气的混合物中混进了空气,在实验过程中发生了爆炸,在没有查明发生事故原因的情况下,就放弃了这项实验。德国化学家能斯特(Nernst,1864—1941),对于氮、氢、氨的气体反应体系也极有兴趣,但由于他在计算时,用了一个错误的热力学数据,以致得出不正确的结论,因而其认为研究这一反应没有前途,停止了该研究。另一位德国的物理学家、化工专家哈伯(Haber,1868—1934)和他的学生一直坚持研究。最初他们尝试在常温下使氮和氢反应,但没有氨气产生,随后在氮、氢混合气中通以电火花,虽有极少量的氨气生成,但耗电量很大。后来他们把注意力集中在高压这个条件上,根据理论计算,氢气和氮气在600 ℃和20 MPa下进行反应,大约生成6%的氨气。如果在高压下将反应进行循环加工,同时不断地分离出生成的氨气,这时就需要催化剂来加快反应速率,从而提高产量。为了探索有效的催化剂,他们进行了大量的实验,发现锇和铀具有良好的催化性能,如果在17.5～20 MPa和500～600 ℃的条件下使用催化剂,氮、氢反应能产生高于6%的氨,因此他们在卡尔斯鲁厄大学建立了一个每小时合成80 g氨的试验装置。这一工艺被德国巴登苯胺纯碱公司接受和采用。由于金属锇稀少、价格昂贵,问题又转向寻找合适的催化剂。在德国化学家米塔斯提议下,于1912年用2 500种不同的催化剂进行了6 500次试验,并终于研制成功含有钾、铝氧化物作助催化剂的价廉易得的铁催化剂。而在工业化过程中碰到的一些难题,如高温下氢气对钢材的腐蚀,碳钢制的氨合成反应器寿命仅有80 h,以及合成氨用氮氢混合气的制造方法,都被该公司的工程师博施所解决。此时,德国国王威廉二世准备发动战争,急需大量炸药,而由氨制得的硝酸是生产炸药的理想原料,于是巴登苯胺纯碱公司于1912年在德国奥堡建成世界上第一座日产30 t合成氨的装置,1913年9月9日开始运转,氨产量很快达到了设计能力。人们称这种合成氨法为哈伯-博施法,它标志着工业上实现高压催化反应的第一个里程碑。由于哈伯和博施的突出贡献,他们分别获得1918年和1931年诺贝尔化学奖。其他国家根据德国发表的论文也进行了研究,并在哈伯-博施法的基础上作了一些改进,先

后开发了合成压力从低压到高压的很多其他方法,这里就不再叙述。

人工合成氨实验的成功令人欢欣鼓舞,它对工业、农业生产和国际科技的重大意义是不言而喻的。但对勒沙特利、能斯特和哈伯三位杰出的科学家而言则是几家欢喜几家愁了。勒沙特利和能斯特的失误则给人们有益的启示,即搞学问要有锲而不舍的精神、坚韧不拔的毅力和一丝不苟的工作作风。但是大家也要记住,勒沙特利和能斯特都是伟大的科学家,虽然在合成氨的研究上没有成功,但是他们在其他领域都有骄人的成就。

3. 氨基甲酸铵的制备

氨基甲酸铵为化学工业中尿素生产过程的生成物,被加热可生成尿素,可燃烧放出有毒氮氧化物和氨气。氨基甲酸铵是白色正方晶系,柱状、板状或片状结晶性粉末,在干燥空气中稳定,但在湿空气中则放出氨而变成碳酸氢铵,在室温下略有挥发,59 ℃时分解为氨及二氧化碳;在密封管中加热至 120~140 ℃时,则失去水变为尿素。其溶解度在 100 g 水中为 66.6 g,能溶于乙醇。市售商品碳酸铵实际就是本品与碳酸氢铵的复盐。

将氨甲基酸铵水溶液放置时,与水反应生成碳酸铵,在酸性溶液中迅速分解。带压操作状态下温度较低时,氨和二氧化碳会反应生成氨基甲酸铵(简称甲铵),反应方程式如下

$$2NH_3(g) + CO_2(g) \rightleftharpoons NH_2COONH_4(s) + 热量$$

该物质是用二氧化碳和氨制脲的中间产物。

氨基甲酸铵转化成尿素的反应是不完全的,需要从含有尿素、过量氨和水的混合物溶液中分离出去。氨基甲酸铵是一种强腐蚀性介质,对使用的材料提出了相应的要求。

相关的反应:

(1)氨基甲酸铵不稳定,易分解为氨和二氧化碳:

$$NH_2COONH_4(s) \longrightarrow 2NH_3(g) + CO_2(g)$$

(2)氨基甲酸铵不稳定,易吸水,溶于水后生成碳酸氢铵和一水合氨:

$$NH_2COONH_4 + 2H_2O \longrightarrow NH_4HCO_3 + NH_3 \cdot H_2O$$

(3)氨基甲酸铵高温下反应生成尿素,这也是工业制备尿素的反应式:

$$NH_2COONH_4 \xrightarrow{\triangle} CO(NH_2)_2 + H_2O$$

(4)氨和二氧化碳气相反应可制备氨基甲酸铵:

$$2NH_3(g) + CO_2(g) \rightleftharpoons NH_2COONH_4 + 热量$$

具体的制备过程是在 1 L 未镀银的硬质玻璃杜瓦瓶中,放入约 400 mL 无水液态氨。塞上带一根弯曲毛细管的塞子,以防止水蒸气在液氨表面上凝聚,毛细管用作氨蒸气的凝气装置。将干冰(固体二氧化碳)打成碎末,慢慢加至液态氨中,继续加入干冰至混合物呈半融的雪浆状。蒸去过量的氨,氨基甲酸铵即留存为块状物。将其转移至真空干燥器中,在略为减压条件下保存 24 h,待残留的氨逸散和少量氨基甲酸铵分解后,即转变为粉末状固体。用 400 mL 液态氨可制得 200~300 g 氨基甲酸铵。

Experiment 4　Determination of Equilibrium Constants for Ammonium Carbamate Decomposition Reaction

【Purpose of the Experiment】

1. Determination of decomposition pressure of ammonium carbamate by isobaric method. Calculating the equilibrium constants and the related thermodynamic functions of decomposition reactions.

2. Master vacuum experiment technology.

【Preview Requirement】

1. Understand the definition of the determination of the equilibrium constant.
2. Understand the method of determining the equilibrium constant.

【Principle】

Ammonium carbamate is an intermediate product of synthetic urea, which is a white solid, very unstable and easily decomposed. At a certain temperature, its decomposition equilibrium reaction is:

$$NH_2COONH_4(s) \rightleftharpoons 2NH_3(g) + CO_2(g)$$

If the gas is treated as an ideal gas, under normal pressure, its equilibrium constant can be approximately expressed as:

$$K^\ominus = \left[\frac{p_{NH_3}}{p^\ominus}\right]^2 \left[\frac{p_{CO_2}}{p^\ominus}\right] \quad (2-4-1)$$

p_{NH_3} and p_{CO_2} are the partial pressures of NH_3 and CO_2 when the reaction reaches the equilibrium. If $p = p_{NH_3} + p_{CO_2}$,

$$p_{NH_3} = \frac{2}{3}p, \quad p_{CO_2} = \frac{1}{3}p \quad (2-4-2)$$

Substitute Equation (2-4-2) into Equation (2-4-1):

$$K^\ominus = \left(\frac{2p}{3p^\ominus}\right)^2 \left(\frac{p}{3p^\ominus}\right) = \frac{4}{27}\left(\frac{p}{p^\ominus}\right)^3 \quad (2-4-3)$$

When the system reaches equilibrium, the equilibrium constant $K^\ominus$ can be calculated by measuring the total pressure p.

The effect of temperature on the equilibrium constant can be expressed as follows:

$$\ln K^{\ominus} = -\frac{\Delta_r H_m^{\ominus}}{RT} + C' \qquad (2-4-4)$$

A plot of $K^{\ominus}$ vs. $1/T$ should yield a straight line with a slope of $-\Delta_r H_m^{\ominus}/R$, and $\Delta_r H_m^{\ominus}$ can be calculated from the slope.

$\Delta_r G_m^{\ominus}$ and $\Delta_r S_m^{\ominus}$ can be calculated by the following equations

$$\Delta_r G_m^{\ominus} = -RT\ln K_p^{\ominus} \qquad (2-4-5)$$

$$\Delta_r S_m^{\ominus} = \frac{\Delta_r H_m^{\ominus} - \Delta_r G_m^{\ominus}}{T} \qquad (2-4-6)$$

Therefore, by measuring the decompression pressure (equilibrium total pressure) of ammonium carbamate at a certain temperature, the above formula can be used to calculate the $K^{\ominus}$, $\Delta_r H_m^{\ominus}$, $\Delta_r G_m^{\ominus}$ and $\Delta_r S_m^{\ominus}$.

【Apparatus and Reagents】

Instrument: 1 set of decomposing pressure measuring instrument (low vacuum digital pressure measuring instrument, isobaric meter, buffer pressure tank); 1 set of constant temperature tank; 1 low vacuum digital pressure gauge; 1 vacuum pump.

Reagent: ammonium carbamate (AR), liquid paraffin.

【Procedures】

1. Record the atmospheric pressure and the room temperature

Record the atmospheric pressure (p_0) and the room temperature by the as-calculated average values before and after experiments.

2. Sample loading

As shown in Figure 2-4-1, put the ammonium carbamate powder (about 0.5 g) into the blue ball from the marked "A" tube nozzle, and then add liquid paraffin to the "U"-shaped tube from the marked "C" tube nozzle. Fill about half the height of the "U"-shaped tube by paraffin used as a liquid seal. Finally, connect the isobaric tube to the vacuum system and seal it.

3. The pressure gauge "takes zero"

As shown in Figure 2-4-1, connect the measuring instrument, turn on the low vacuum digital pressure gauge, warm up the power supply for 2 min, switch the pressure display unit of the differential pressure gauge to "kPa", open the intake valve, and wait for the pressure gauge to display a stable number, press the "Zero" button to eliminate the zero drift of the instrument system, finally, display "0000" on the LED.

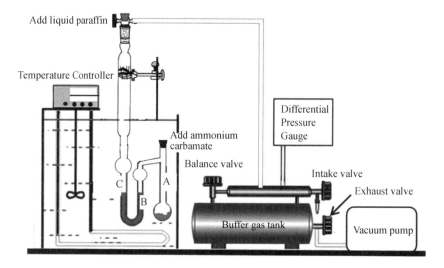

Figure 2 – 4 – 1 Measuring device for ammonium carbamate decomposition equilibrium constant

4. Check the air tightness of the device

(1) Check the overall air tightness of the buffer gas storage tank

Tighten the rubber pipe clamp in Figure 2 – 4 – 1, then close the intake valve, open the balance valve and the exhaust valve (all three valves are closed clockwise, turn counterclockwise). Start the vacuum pump until the pressure gauge number is 100 kPa ~ 200 kPa, close the exhaust valve, stop the vacuum pump, and observe the change in the value of the digital pressure gauge. If the increase rate of the pressure gauge reading is less than 0.01 kPa/s, the overall air tightness is good. Otherwise, it is necessary to find and remove the cause of air leakage until the air tightness is good.

If it is judged that the air is leaking at this time, close the balance valve, and the pressure gauge reading increase rate is greater than 0.02 kPa/s, then the intake valve is likely to leak. On the contrary, the air extraction valve or balance valve is leaking.

(2) Air tightness inspection of the system under test

After confirming that the buffer gas storage tank is not leaking, turn on the rubber hose clamp switch to connect the system under test with the buffer gas storage tank. Observe that the increased rate of the pressure is less than 0.01 kPa/s, indicating that the air tightness is good. Otherwise, check the rubber tube and each interface until the air tightness is qualified. After the leak detection is completed, open the intake valve to relieve the pressure of the fine-tuning part to zero.

5. Determination of decomposition pressure

Adjust the temperature of the thermostatic bath until the temperature of the water bath is constant to 35 ℃, close the air inlet valve, open the air extraction valve and the balance valve, start the vacuum pump to pump the system. When the pressure decreases from 0 to about

−90 kPa, close the intake valve, and then turn off the vacuum pump. At this time, the liquid level of the two pipes in the "U"-shaped tube is not at the same level. The liquid level of C tube is obviously higher than that of B tube. Slowly open the intake valve to increase the system pressure to avoid the liquid in the B tube poured back into the A tube. The intake valve is continuously adjusted to keep the same level of the liquids in B and C tubes for more than 2 min, indicating that the ammonium carbamate in A tube has been close to the decomposition equilibrium. Record the quasi-equilibrium pressure. It should be noted that the as-recorded pressure is not the decomposition pressure of ammonium carbamate, and the decomposition of ammonium carbamate can be obtained by further calculating from the the as-recorded pressure.

Turn on the suction pump again to reduce the pressure of the system, repeat the operation in step 5, measure the parallel data, and require that the pressure difference between the two measurements is less than 0.3 kPa, record the pressure displayed from the pressure gauge, and calculate the average.

Continue to determine the decomposition pressure at 30 ℃, 35 ℃, 40 ℃, 45 ℃ and 50 ℃ in accordance with step 5.

6. End of experiment

Slowly open the intake valve (to prevent the liquid in B tube from being poured into A tube), put air in to relieve the pressure of the tested system to zero (slowly close the balance valve, open the intake valve), remove the isobaric pipe, and turn off all power, and organize the experimental table.

【Records and Analysis of Data】

1. Record the experimental data into the table 2 − 4 − 1.

Room temperature:_____   atmospheric pressure:_____

Table 2 − 4 − 1 Experimental data of decompression and equilibrium constant

$t/℃$	$1/T$ $\times 10^3/K^{-1}$	Vacuum degree $\Delta P/kPa$	Decompression $p = p_0 - |\Delta p|$	$K^\ominus$	$\ln K_p^\ominus$

2. Make ln $K^{\ominus}$ ~ 1/T diagram, and calculate the $\Delta_r H_m^{\ominus}$ based on the slope of this line.

3. Calculate the $\Delta_r G_m^{\ominus}$ and $\Delta_r S_m^{\ominus}$ according to the relevant thermodynamic formula. The thermodynamic data required in the experiment is referred to the table.

4. The relative error of ammonium carbamate decompression measured by experiment is calculated by comparing with the data in the table.

【Notes】

1. The vacuum pump should connect the air inlet valve of the pump with the atmosphere before stopping the pump to prevent the oil backdraft pollution test system.

2. During the experiment, the inlet valve and balance valve must be opened slowly, and the inlet valve and balance valve must be closed, which is one of the important operations to ensure the smooth test.

3. When the decomposition pressure of several temperatures is measured after the experiment, it is not necessary to re-pump the gas, and only the liquid level is adjusted.

4. The device used in this experiment is similar to the device used to measure the saturated vapor pressure of liquid, so it can be used to measure the saturated vapor pressure of liquid.

【Post Lab Questions】

1. What is decomposition pressure? How does the amount of ammonium carbamate affect the decomposition pressure at a certain temperature?

2. How to determine the decomposition pressure of ammonium carbamate? Does the number of samples have an effect on the total equilibrium pressure?

3. How to judge the decomposition of ammonium carbamate has reached equilibrium? If there is no balance, what will be the impact on the measured data?

4. What happens if there is too much ventilation when air is introduced into the system? How to overcome it?

实验5　乙酸乙酯皂化反应速率常数的测定

【实验目的】

1. 了解电导法测定乙酸乙酯皂化反应的速率常数和活化能；
2. 了解二级反应的特点，学会用图解法求二级反应的速率常数；
3. 熟悉电导率仪的使用方法。

【预习要求】

1. 了解电导法测定化学反应速率常数和活化能的原理；

2. 了解电导率仪的使用方法及铂黑电极的使用与保管方法；
3. 了解该实验的注意事项。

【实验原理】

乙酸乙酯皂化反应是一个典型的二级反应,其反应速率与乙酸乙酯和碱的浓度乘积成正比。

设起始的乙酸乙酯浓度和碱的起始浓度相同,并以 C_0 表示,而当反应进行到某一时间 t 时,两者浓度减少均为 x,即

$$\text{CH}_3\text{COOC}_2\text{H}_5 + \text{NaOH} \longrightarrow \text{CH}_3\text{COONa} + \text{C}_2\text{H}_5\text{OH}$$

$t=0$ 时的浓度 　　C_0 　　　C_0 　　　0 　　　0

$t=t$ 时的浓度 　　C_0-x 　C_0-x 　x 　　x

则此二级反应速率方程为

$$\frac{dx}{dt} = k(C_0 - x)^2 \tag{2-5-1}$$

式中,k 为反应速率常数。

将式(2-5-1)积分得

$$\int_0^x \frac{dx}{(C_0 - x)^2} = \int_0^t k\,dt \tag{2-5-2}$$

得到动力学方程为

$$\frac{x}{C_0(C_0 - x)} = kt \tag{2-5-3}$$

如果用化学方法,测出不同时间碱的浓度,依式(2-5-3)按作图法可以求出 k 值,但操作比较麻烦。如果采用直接测量反应体系电导率的方法,操作就简便得多,同样可以达到精确测定皂化反应速率常数 k 的目的。

首先,假定反应是在极稀的水溶液中进行的,离子的电导率与离子的浓度成正比。在反应体系中,参加电导的只有 Na^+、OH^- 和 CH_3COO^- 三种离子。而 Na^+ 在反应前后的浓度实际是不变的。随反应的进行,OH^- 不断减少,CH_3COO^- 则不断增加,但 OH^- 的迁移率比 CH_3COO^- 的迁移率大得多。所以,致使反应体系的电导率随时间的推移不断下降。

显然,电导率的减少值是和 OH^- 浓度的减少值 x 成正比的,即在 $t=0$ 时,有

$$x = A(L_0 - L_t) \tag{2-5-4}$$

式中,A 为比例常数;L_0 为起始溶液(浓度为 C_0 的 NaOH 溶液)的电导率;L_t 为 t 时溶液的电导率。

将式(2-5-4)代入式(2-5-3)得

$$kt = \frac{A(L_0 - L_t)}{C_0[C_0 - A(L_0 - L_t)]} \tag{2-5-5}$$

将等式两端颠倒得

$$\frac{1}{kt} = \frac{C_0^2}{A(L_0 - L_t)} - C_0 \tag{2-5-6}$$

整理得

$$\frac{1}{L_0 - L_t} = \frac{A}{C_0^2 kt} + \frac{A}{C_0} \qquad (2-5-7)$$

令

$$B = \frac{A}{C_0}, \quad m = \frac{A}{C_0^2 k} = \frac{B}{C_0 k} \qquad (2-5-8)$$

则得

$$\frac{1}{L_0 - L_t} = m \frac{1}{t} + B \qquad (2-5-9)$$

式(2-5-9)表明,若以 $\frac{1}{L_0 - L_t}$ 对 $\frac{1}{t}$ 作图,应得一直线,由直线的斜率 m 与截距 B 的数值即可求出反应速率常数 k。

反应的活化能可根据 Arrhenius 公式求得

$$\frac{d\ln k}{dT} = \frac{E_a}{RT^2} \qquad (2-5-10)$$

积分得

$$\ln \frac{k(T_2)}{k(T_1)} = \frac{E_a(T_2 - T_1)}{RT_1 T_2} \qquad (2-5-11)$$

式中,$k(T_1)$、$k(T_2)$ 分别对应温度为 T_1、T_2 的反应速率;R 为摩尔气体常数;E_a 为反应的活化能。

【仪器与试剂】

恒温水浴 1 套,电导率仪(DDS-307A)1 台,秒表 1 块,电导瓶(100 mL 锥形瓶 1 个),50 mL 移液管(或 50 mL 碱式滴定管)1 支,100 μL 注射器 1 支,0.010 0 mol·L^{-1} NaOH 溶液,乙酸乙酯(分析纯),滤纸。

【实验步骤】

1. 开启电导率仪的电源,预热 10 min,调节电导率仪(参阅仪器介绍)。

2. 温度 T_1 时 L_0 的测定。

L_0 即反应体系不加入乙酸乙酯时的 NaOH 溶液的电导率。本实验要求 NaOH 的浓度与乙酸乙酯浓度相等。故根据下面公式可计算需加 NaOH 准备液的体积 $V_{碱}$

$$V_{碱} = \frac{d_{酯} \cdot V_{酯} \cdot 1\,000}{M_{酯} \cdot C_0} \quad (\text{mL})$$

式中,$V_{酯}$ 为需加乙酸乙酯的体积(0.080 mL);$d_{酯}$ 为室温下乙酸乙酯的密度(参见本书附录中附表-6 和附表-17);$V_{碱}$ 为所加碱的体积;$M_{酯}$ 为乙酸乙酯相对分子质量(88.11);C_0 为 NaOH 准备液的浓度,取 0.010 0 mol·L^{-1}。

按计算的数据,用移液管或碱式滴定管准确量取体积为 $V_{碱}$ 的 0.010 0 mol·L^{-1} 的 NaOH 溶液于 100 mL 洁净干燥的锥形瓶中,放在恒温水浴中(温度为 T_1)。用蒸馏水轻轻淋洗铂黑电极3次,再用滤纸吸干电极表面的水分(切勿触及铂黑)。将电极插入电导瓶(锥形瓶)中并与电导率仪接好,待恒温10 min后,测定其电导率,直至稳定不变为止,即为温度T_1时的L_0。

3. 温度T_1时L_t的测定。

用微量注射器取0.080 mL的乙酸乙酯于第2步中的电导瓶中,同时开动停表,作为反应的起始时间(注意停表一经打开切勿按停,直至全部实验结束)。然后小心地将溶液摇匀,记录反应进行到表2-5-1所列时刻溶液的电导率(L_t)。实验完毕,将铂黑电极取出用蒸馏水淋洗干净并浸泡在蒸馏水中,电导瓶洗净烘干。

4. 调节恒温水浴,使温度为T_2($T_2 = T_1 + 10 ℃$),重复上述步骤测T_2时的L_0和L_t。

5. 实验完成后,将铂黑电极取出用蒸馏水淋洗干净并浸泡在蒸馏水中,电导瓶洗净烘干。

【实验记录及数据处理】

1. 反应温度:$T_1 = $ _____℃, $T_2 = $ _____℃

反应物起始浓度 $C_0 = $ _____

$T_1 = $ _____℃ $T_2 = $ _____℃

表2-5-1 乙酸乙酯皂化反应过程电导实验数据

t/min	$1/t$	L_t		$L_0 - L_t$		$1/(L_0 - L_t)$	
		T_1	T_2	T_1	T_2	T_1	T_2
2							
4							
6							
8							
10							
13							
16							
19							
21							
25							
30							

2. 分别以T_1和T_2时的$1/t$对$1/(L_0 - L_t)$作图,求其斜率并由此计算室温和35 ℃时的k值。

3. 分别由T_1和T_2所求的k值,按Arrhenius公式求出反应的活化能E_a。

4. 文献值见表2-5-2。

表 2-5-2 乙酸乙酯皂化反应速率常数及活化能文献值

$c(CH_3COOC_2H_5)$ /(mol·L^{-1})	$c(OH^-)$ /(mol·L^{-1})	t/K	k /(L·mol^{-1}·s^{-1})	k /(L·mol^{-1}·min^{-1})	E /(kcal·mol^{-1})①
0.01	0.02	273.15	8.60×10^{-3}	0.519	14.6
		283.15	2.35×10^{-2}	1.410	
		292.15	5.03×10^{-2}	3.020	
0.021	0.023	298.15		6.850	
$\lg k = -1780/T + 0.00754 + 4.53$					

注:① 1 kcal = 4.186 8 kJ。

【讨论与说明】

1. 本实验所用的蒸馏水需事先煮沸,待冷却后使用,以免溶有的 CO_2 与溶液中的 NaOH 发生反应,降低 NaOH 溶液的浓度。

2. 清洗铂电极时不可用滤纸擦拭电极上的铂黑。

3. 如果恒温水浴的温度波动超过 ±0.05 ℃范围,会对皂化反应的速率与作图时的线性产生较大影响。

4. 停表要连续计时,不能中途停止。

5. 所用 NaOH 溶液和乙酸乙酯的浓度必须相等。

6. 动力学实验,一般情况下要在恒温条件下进行,因为不同温度下反应速率不一样,反应速率常数与温度有关,电导率的数值也与温度有关,所以要准确进行动力学测量,必须在恒温条件下进行。

7. 乙酸乙酯皂化反应为吸热反应,混合后体系温度降低,所以在混合后的起始几分钟内所测溶液的电导率偏低,因此最好在反应 4~6 min 后开始,否则 L_t 对 $(L_0-L_t)/t$ 作图得到的是一条抛物线,而不是直线。

8. 乙酸乙酯皂化反应曲线随着时间的延长,会出现偏离二级反应的现象。对此,有人认为"皂化反应曲线是双分子反应"的说法欠妥,该反应是一种"表观二级反应";随着反应时间的延长,反应的可逆性对总反应的影响逐渐变得明显。又有人认为,皂化反应中还存在盐效应,即某些中性盐的存在会降低其速率系数。因此,皂化反应实验的时间以半小时为宜,至多不超过 40 min。

9. 求反应速率的方法主要有物理化学分析法和化学分析法两大类。物理化学分析法有旋光、折光、电导、分光光度等方法,根据不同情况可用不同仪器。这些方法的优点是实验时间短、速度快,可不中断反应,而且还可采用自动化的装置,但是需相关的仪器设备,并且只能得出间接的数据,有时往往会因某些杂质的存在而产生较大的误差。化学分析法是在一定时间内取出一部分试样,使用预冷或去催化剂等方法使反应停止,然后进行分析,直接求出浓度。

【思考题】

1. 如果反应物起始浓度不相等,对本实验结果有无影响?
2. 为什么实验要在恒温的条件下进行?
3. 实验过程中能否换电极?

【相关阅读】

电导、电导率的测量及其应用

电导是电化学中的一个重要参量,它不仅反映电解质溶液中离子状态及其运动的许多信息,而且由于它在稀溶液中与离子浓度之间的简单线性关系,被广泛应用于分析化学和动力学过程的测试中。

电导值是电阻值的倒数。因此,电导值实际上是通过电阻的测量,然后计算电阻的倒数求得的。电解质溶液电导的测量本身有其特殊性,因为溶液中离子导电机理与金属中电子的导电机理不同。伴随电导过程,离子在电极上放电,因而会使电极发生极化现象。因此,溶液电导值的测量通常都是用较高频率的交流电桥来实现的,大多数电导测量所用的电极均镀以铂黑来减少电极本身的极化作用。

溶液电导的测量是通过一对金属电极组成的电导池进行的。当温度一定时,被测溶液呈现在测量电极之间的电导值 G 与溶液电导率 κ 及电极面积 A 成正比,与两个电极的距离 l 成反比,即

$$G = \kappa \frac{A}{l}$$

定义测量电极相隔的距离和电极面积的比值(l/A)为电导池常数,单位为 m^{-1}。电导池常数是一个电导池特征值,但要精确测定电导池中的 l 与 A 值是很困难的,一般用间接的方法来测定。将一已知电导率的标准溶液(通常用一定浓度的 KCl 溶液)装入电导池中,在指定温度下,测其电导值 G,再根据 $G = \kappa(A/l)$ 求算电导池常数。

电导值的单位是西门子(S),电导率的单位则是 $S \cdot m^{-1}$。

如果把含有 1 mol 电解质的溶液置于相距为 1 m 的电导池的两极之间,这时所具有的电导为摩尔电导率 Λ_m。若电解质溶液的浓度为 $C(mol \cdot L^{-1})$,则 Λ_m 与 C 的关系为

$$\Lambda_m = \frac{\kappa}{C}$$

因此,测定一定浓度的 KCl 水溶液的摩尔电导率 Λ_m,并查得该浓度下的 κ 值(参阅附录),也可求得电导池常数。

电导及电导率可用于弱电解质电离常数的测定、难溶盐的溶度积测定等。

Experiment 5 Determination of Rate Constant for the Saponification of Ethyl Acetate

【Purpose of the Experiment】

1. To know the rate constant and activation energy for ethyl acetate saponification.
2. To know the characteristics of the second-order reaction and learn the graphical method to determine the rate constant of the second-order reaction.
3. To be familiarized with the operation of a conductometer.

【Principle】

The saponification of ethyl acetate in sodium hydroxide solution is a typical second-order, in which the reaction rate is proportional to the concentration of ethyl acetate and the alkali.

The concentration of the initial concentrations of ethyl acetate and alkali is the same, named as C_0. When the reaction time is t, the concentrations of the produced sodium acetate and ethanol are all recorded as x, then the concentration of ethyl acetate and alkali is $C_0 - x$.

$$CH_3COOHC_2H_5 + NaOH \longrightarrow CH_3COONa + C_2H_5OH$$

$t = 0$	C_0	C_0	0	0
$t = t$	$C_0 - x$	$C_0 - x$	x	x

Then the rate equation for this second-order reaction can be expressed as:

$$\frac{dx}{dt} = k(C_0 - x)^2 \qquad (2-5-1)$$

Where k is rate constant.
Integrating the Equation (2-5-1):

$$\int_0^x \frac{dx}{(C_0 - x)^2} = \int_0^t k dt \qquad (2-5-2)$$

A new kinetic equation is obtained as Equation (2-5-3):

$$\frac{x}{C_0(C_0 - x)} = kt \qquad (2-5-3)$$

Obviously, if the concentration of alkali or ethyl acetate at different times t can be measured, the $x - t$ curve can be drawn by Equation (2-5-3) to obtain the k value. However, x is difficult to be determined by chemical methods. Fortunately, the concentration of hydroxide is approximately proportional to the conductivity of the solution. Measuring the conductivity of a reaction system is simple and accurate.

Due to the low concentration of the reactants, the conductivity of the ions is proportional to the concentration of the ions. The conductivity of the reaction solution is derived from sodium ions, hydroxide ions, and acetate ions. The sodium ion concentration remained constant throughout the reaction, indicating that sodium ions have no effect on conductivity changes.

The effect of acetate ions for conductivity can be ignored since the molar conductivity of the hydroxide ion is much larger than that of the acetate ions. Therefore, the effect of slight changes in acetate ions for the conductivity is also negligible. With the increase of the reaction time, the hydroxide ion concentration is continuously decreased and the acetate concentration is gradually increased, however, the decrease in the solution conductivity is only relative to the as-reduced concentration of hydroxide ion. So,

$$x = A(L_0 - L_t) \quad (2-5-4)$$

Where A is the proportionality constant, L_0 and L_t is the conductivities at $t = 0$ (C_0 of NaOH solution) and $t = t$, respectively. Substituting Equation (2-5-4) into Equation (2-5-3), and then to obtain Equation (2-5-5).

$$\frac{1}{L_0 - L_t} = \frac{A}{C_0^2 kt} + \frac{A}{C_0} \quad (2-5-5)$$

From the slope and intercept of the fitted straight line by the Equation (2-5-5), the reaction rate constant k can be calculated.

The activation energy can be calculated by the Arrhenius Equation (2-5-6).

$$\ln \frac{k_2}{k_1} = \frac{E_a(T_2 - T_1)}{RT_2T_1} \quad (2-5-6)$$

Where k_1 and k_2 represent the rate constants in the reaction temperatures of T_1 and T_2, respectively. E_a is the activation energy. R is gas constant.

【Apparatus and Reagents】

Constant temperature water bath; Conductivity meter (DDS-307A); Stopwatch; Conductivity bottle; 50 mL of basic burette; 100 microliter syringe; 0.010 0 mol·L^{-1} NaOH; Ethyl acetate (analytical grade) and Filter paper.

【Procedures】

1. Turn on the power of the conductivity meter and warm up for 10 min. Adjust the bath temperature at 25 ℃.

2. Determination of L_0 at 25 ℃.

The required sodium hydroxide volume can be calculated by Equation (2-5-7).

$$V = \frac{d \cdot 0.08 \cdot 1\,000}{M \cdot C_0} \quad (2-5-7)$$

Where V is the volume of 0.010 0 mol·L^{-1} NaOH. d is the density of ethyl acetate at 25 ℃, and M is the molar mass of ethyl acetate (88.11).

Subsequently, V of a 0.010 0 mol·L^{-1} NaOH solution was added to a 100 mL clean and dry conical flask by an alkaline burette. Then place the conical flask in a water bath at 25 ℃ to constant temperature. A platinum black electrode is gently rinsed with distilled water, dried by using the filter paper to absorb the distilled water on the surface of the electrode (Don't touch the platinum black), finally inserted into the conical flask containing 0.010 0 mol·L^{-1} NaOH solution. After the conductivity is constant, the value is the L_0 value at 25 ℃, and the value is recorded in Table 2-5-1.

3. Determination of L_t at temperature 25 ℃

Pipette 0.080 mL of ethyl acetate with a micro-syringe and quickly add it to the thermostated conical flask. Simultaneously, press the stopwatch and quickly shake the solution to make it homogeneous. It must be noted that the stopwatch continues until the end of the experiment. In the experiment, shake carefully the solution and record the conductivity (L_t) of the solution at the corresponding time in Table 2-5-1. After the experiment was completed, the platinum black electrode was taken out, rinsed with distilled water and immersed in distilled water, and then the conductivity bottle was washed and dried.

4. The thermostatic water bath was adjusted to 30 ℃, and the above procedures were repeated to measure the L_0 and L_t at 30 ℃.

【Records of Data】

Reaction temperature and initial conductivity:

$$T_1 = 25\ ℃, L_0 = \underline{\qquad};$$
$$T_2 = 30\ ℃, L_0 = \underline{\qquad}.$$

Table 2-5-1 Experimental data of ethyl acetate saponification

t/min	L_t		$L_0 - L_t$		$1/(L_0 - L_t)$	
	25 ℃	30 ℃	25 ℃	30 ℃	25 ℃	30 ℃
2						
4						
6						
8						
10						
13						
16						
19						
21						
25						
30						

【Data Analysis】

1. Plot the $\dfrac{1}{L_0 - L_t} \sim \dfrac{1}{t}$ curve. Calculation the reaction rate constants (k_1 and k_2 at 25 ℃

and 30 ℃, respectively) from the slopes and intercepts of the plotted curves.

2. Calculate activation energy of the saponification of ethyl acetate.

【Notes】

1. The reported values of rate constant and activation energy for cthyl acetate saponification in Table 2 – 5 – 2.

Table 2 – 5 – 2 The literature values of the rate constant and the activation energy of ethyl acetate saponification reaction

$c(CH_3COOC_2H_5)$ /(mol·L^{-1})	$c(OH^-)$ /(mol·L^{-1})	t/K	k /(L·mol^{-1}·s^{-1})	k /(L·mol^{-1}·min^{-1})	E /(kcal·mol^{-1})[①]
0.01	0.02	273.15	8.60×10^{-3}	0.519	14.6
		283.15	2.35×10^{-2}	1.410	
		292.15	5.03×10^{-2}	3.020	
0.021	0.023	298.15		6.850	
$\lg k = -1\,780/T + 0.007\,54 + 4.53$					

Note：① 1 kcal = 4.186 8 kJ。

2. The concentrations of NaOH solution and ethyl acetate must be equal.

3. The platinum black on the electrode should not be wiped off by a filter paper.

【Post Lab Questions】

1. Whether the electrode can be replaced during the experiment?

2. Why is the experiment carried out under constant temperature conditions?

实验6 蔗糖水解反应速率常数的测定

【实验目的】

1. 用旋光仪测定蔗糖溶液水解时的旋光度随时间的变化规律,从而推算蔗糖水解反应的速率常数和半衰期;

2. 了解旋光仪的基本原理,掌握旋光仪的正确使用方法;

3. 了解催化剂对反应速率和反应速率常数的影响。

【预习要求】

1. 了解旋光仪测定蔗糖水解速率常数的原理和方法;

2. 了解和熟悉旋光仪的构造、原理和使用方法。

【实验原理】

蔗糖在水中转化成葡萄糖与果糖的反应为

$$C_{12}H_{22}O_{11} + H_2O \xrightarrow{H^+} C_6H_{12}O_6 + C_6H_{12}O_6 \quad (2-6-1)$$
$$\text{蔗糖} \qquad\qquad\qquad \text{葡萄糖} \quad \text{果糖}$$

在纯水中，此反应的速率极慢。为使水解反应加速，反应常以 H_3O^+ 为催化剂，故该反应在酸性介质中进行。此反应本为一个二级反应，但由于蔗糖水溶液较稀，反应时水是大量存在的，尽管有部分水分子参加了反应，仍可近似认为整个反应过程中水的浓度基本上保持不变。因此，蔗糖水解反应可看作一级反应（确切地说为"准一级反应"）。

一级反应速率方程为

$$-\frac{dC}{dt} = kC \quad (2-6-2)$$

积分得

$$t = \frac{1}{k}\ln\frac{C_0}{C} \quad (2-6-3)$$

式中，k 为反应速率常数；C 为反应时间为 t 时反应物的浓度；C_0 为反应开始时反应物的浓度。

当 $C = \frac{1}{2}C_0$ 时，时间 t 可用 $t_{1/2}$ 表示，即反应的半衰期，则

$$t_{1/2} = \frac{\ln 2}{k} = \frac{0.6931}{k} \quad (2-6-4)$$

可见一级反应的半衰期只决定于反应速率常数 k，而与反应物起始浓度无关。若测得反应在不同时刻时蔗糖的浓度，代入上述动力学的公式中，即可求出 k 和 $t_{1/2}$。

测定反应物在不同时刻的浓度可用化学法和物理法，本实验采用物理法测定反应系统旋光度的变化。

在本实验中，蔗糖及其转化产物葡萄糖与果糖都含有不对称的碳原子，它们都具有旋光性，而且它们的旋光能力不同，故可以利用体系在反应过程中旋光度的变化来度量反应进程。

测量物质旋光度所用仪器称为旋光仪。溶液的旋光度与溶液中所含旋光物质的旋光能力、溶剂性质、溶液浓度、样品长度及温度等均有关系，当其他条件均固定时，旋光度 α 与反应物浓度 C 呈线性关系，即

$$\alpha = BC \quad (2-6-5)$$

式中，比例常数 B 与物质的旋光能力、溶剂性质、光源的波长、样品管长度、反应时的温度等有关。

物质的旋光能力用比旋光度来度量。比旋光度可用下式表示

$$[\alpha]_D^{20} = \frac{100\alpha}{lC} \quad (2-6-6)$$

式中，$[\alpha]_D^{20}$ 右上角的 20 表示实验时温度为 20 ℃；D 是指所用光源为钠光灯光源 D 线；α 为测得的旋光度，(°)；l 为样品管长度，dm；C 为浓度，g/100 mL。

作为反应物的蔗糖是右旋性物质，其比旋光度 $[\alpha]_D^{20} = 66.6°$；生成物中葡萄糖也是右旋性物质，其比旋光度 $[\alpha]_D^{20} = 52.5°$；但果糖却是左旋性物质，其比旋光度 $[\alpha]_D^{20} = -91.9°$。因此，随着反应的进行，物质的右旋角不断减小，反应至某一瞬间，物系的旋光度恰好等于零，随后就变为左旋，直至蔗糖完全转化，这时左旋角达到最大值 α_∞。

设反应开始时体系的旋光度为 α_0，则

$$\alpha_0 = B_\text{反} C_0 \quad (t = 0 \text{ 时}, \text{蔗糖尚未转化}) \quad (2-6-7)$$

反应终了时体系的旋光度为 α_∞，则

$$\alpha_\infty = B_\text{产} C_0 \quad (t = \infty \text{ 时}, \text{蔗糖全部转化}) \quad (2-6-8)$$

上述两式中 $B_\text{反}$ 和 $B_\text{产}$ 分别为反应物与生成物的比例常数。

当反应时间为 t 时，蔗糖浓度为 C，此时旋光度 α_t 为

$$\alpha_t = B_\text{反} C + B_\text{产}(C_0 - C) \quad (2-6-9)$$

由式(2-6-7)和式(2-6-9)联立可解得

$$C_0 = \frac{\alpha_0 - \alpha_\infty}{B_\text{反} - B_\text{产}} = B(\alpha_0 - \alpha_\infty) \quad (2-6-10)$$

$$C = \frac{\alpha_t - \alpha_\infty}{B_\text{反} - B_\text{产}} = B(\alpha_t - \alpha_\infty) \quad (2-6-11)$$

将式(2-6-10)和式(2-6-11)代入式(2-6-3)，得

$$t = \frac{1}{k} \ln \frac{C_0}{C} = \frac{1}{k} \ln \frac{\alpha_0 - \alpha_\infty}{\alpha_t - \alpha_\infty} \quad (2-6-12)$$

即

$$\lg(\alpha_t - \alpha_\infty) = -\frac{k}{2.303}t + \lg(\alpha_0 - \alpha_\infty) \quad (2-6-13)$$

由式(2-6-13)可以看出，如以 $\lg(\alpha_0 - \alpha_\infty)$ 对 t 作图可得一直线，由直线的斜率即可求得反应的速率常数 k。

本实验使用旋光仪测定蔗糖水解过程中不同时间的旋光度 α_t 以及全部水解后的旋光度 α_∞ 值，通过作图求得速率常数 k。

【仪器与试剂】

旋光仪 1 台，停表 1 块，旋光管(带有恒温水外套) 1 支，恒温水浴 1 套，容量瓶(50 mL)

1个,电子天平(精确到 0.01 g)1 台,锥形瓶(100 mL)2 个,移液管(25 mL)2 支,烧杯 100 mL,500 mL 各 1 个,1 mol·L^{-1}和 2 mol·L^{-1} HCl 溶液。

【实验步骤】

1. 调整恒温水浴至 25 ℃恒温,然后将旋光管外套接上恒温水。
2. 旋光仪零点的校正。

蒸馏水为非旋光物质,可以用来校正旋光仪的零点(即 $\alpha = 0°$时仪器对应的刻度)。用蒸馏水洗净旋光管各部分零件(玻璃片、垫圈要单独拿着洗,以防掉进下水道)。将旋光管一端的盖子旋紧,向管内注满蒸馏水,将玻璃片盖上,旋紧套盖(小心操作,以防用力过猛,压碎玻璃片),然后用滤纸擦干旋光管的外面,最后用镜头纸擦净旋光管两端玻璃片。在放入旋光仪前观察旋光管中是否有气泡,如果有小气泡,应设法将小气泡赶至管的凸肚部分。将旋光管放入旋光仪内,凸肚一端位于上方,盖上槽盖。打开旋光仪电源开关,调节目镜使视野清晰,然后旋转检偏镜,直到在两个三分视野的中间找到明暗相等的视野为止(即观察不到三分视野,见附录)。记下刻度盘的读数,重复操作三次,取其平均值,此即旋光仪的零点。测完后取出旋光管,倒出蒸馏水。

3. 蔗糖水解过程中 α_t 的测定。

称取蔗糖 10 g,溶于蒸馏水中,用 50 mL 容量瓶配制成溶液(如混浊需过滤)。用移液管取 25 mL 蔗糖溶液和 25 mL 2 mol·L^{-1} HCl 溶液分别注入两个 100 mL 干燥锥形瓶中,并将这两个锥形瓶同时置于 25 ℃恒温水浴中恒温 10~15 min。待恒温后,将 25 mL 蔗糖溶液迅速倒入 25 mL HCl 溶液中,并开动停表作为反应的起始时间开始计时。迅速振荡摇动至均匀,取少量混合液清洗旋光管两次,然后将此混合液注满旋光管,盖好玻璃片,旋紧套盖(检查是否漏液和有无气泡)。擦净旋光管及两端玻璃片(放入旋光仪测定时需注意旋光管中是否有气泡,处理方法同步骤2)。测量 t 时刻溶液的旋光度 α_t。测定时快速将溶液置于旋光仪中,不测定时放入 25 ℃恒温水浴中恒温,测定也要迅速(防止温度变化过大)。调好读数位置后,先记下时间,再读取旋光度数值。可在测定第一个 α_t 值后的 5 min、10 min、15 min、20 min、30 min、50 min、75 min、100 min 各测一次,并将测量结果记录于表 2-6-1 中。

4. α_∞ 的测定

要使蔗糖完全水解,通常需 48 h 左右,将步骤 3 中剩余的混合液保留好,48 h 后重新恒温观测其旋光度,此值即为 α_∞。也可将剩余混合液置于 50~60 ℃水浴中温热 40 min,加速水解反应进行,然后冷却至测量温度。按上述操作测其旋光度,即认为是 α_∞。注意水浴温度不可过高,否则将产生副反应,溶液颜色变黄;在加热过程中要盖好瓶塞,防止溶液蒸发影响浓度。

实验结束时,应立即将旋光管洗净擦干,防止酸对旋光管的腐蚀。

5. 选择 1 mol·L^{-1} HCl 溶液为催化剂,其他步骤重复步骤 3 和步骤 4。

【实验记录与数据处理】

1. 将实验数据记入表 2-6-1。

室温：_____℃   大气压：_____ kPa
实验温度：_____℃   旋光管长度：$l =$ _____ dm
旋光仪零点校正值：_____ α_∞：_____

表 2-6-1 蔗糖水解反应实验数据

反应时间 /min	α_t		$\alpha_t - \alpha_\infty$		$\lg(\alpha_t - \alpha_\infty)$	
	$C_{HCl} = 2$ mol·L^{-1}	$C_{HCl} = 1$ mol·L^{-1}	$C_{HCl} = 2$ mol·L^{-1}	$C_{HCl} = 1$ mol·L^{-1}	$C_{HCl} = 2$ mol·L^{-1}	$C_{HCl} = 1$ mol·L^{-1}

2. 以 $\lg(\alpha_t - \alpha_\infty)$ 对 t 作图，由所得直线的斜率求速率常数 k 值，并计算蔗糖水解反应的半衰期 $t_{1/2}$。

3. 比较不同催化剂浓度下反应速率和反应速率常数之间的关系。

【讨论与说明】

1. 蔗糖在配制溶液前，需先经 110 ℃ 烘干。

2. 在进行蔗糖水解反应速率常数的测定以前，需熟练掌握旋光仪的使用，能正确而迅速地读出其读数。

3. 旋光管管盖只要旋至不漏水即可，过紧易造成损坏，或因玻片受力产生应力而致使有一定的假旋光。

4. 旋光仪的钠光灯不宜长时间开启，测到 30 min 后，每次测量间隔时应将钠光灯熄灭，以延长钠光灯寿命。但下一次测量之前应提前 10 min 打开钠光灯,使光源稳定。

5. 由于混合液酸度较大,每次测量时旋光管外面一定要擦净后才能放入旋光仪内。实验结束时,应立即将旋光管洗净擦干,防止酸对旋光管的腐蚀。

6. 实验记录的时间应以调节出视场均匀的时刻为准,实验步骤所给时间仅供参考,提供给大家记录数据的大致时间间隔。

【思考题】

1. 蔗糖水解的速度与哪些因素有关?
2. 为什么可用蒸馏水来校正旋光仪零点?
3. 配制蔗糖溶液时称量不够准确,对测量结果是否有影响?

【相关阅读】

1. 蔗糖水解反应速率的影响因素

蔗糖是从甘蔗内提取的一种纯有机化合物,也是与生活关系最密切的一种天然碳水化合物,它是由 D-(-)-果糖和 D-(+)-葡萄糖通过半缩酮和半缩醛的羟基相结合而生成的。蔗糖经酸性水解后,产生一分子 D-葡萄糖和一分子 D-果糖。

蔗糖水解的反应速率与蔗糖的浓度、水的浓度、催化剂氢离子的浓度及温度有关。为了顺利测定该反应的速率常数,在恒定温度和氢离子浓度的情况下,该反应是一个二级反应,但由于水是大量的,反应前后水的浓度可近似认为是恒定的,因此蔗糖转化反应可看作为准一级反应。由上面的条件可知,反应对氢离子的浓度要求非常严格,小小的误差就会造成实验的条件发生变化;蔗糖浓度虽然对反应有影响,但是微量的误差对实验的测定影响不大,关键是条件的设定。因而配制蔗糖溶液可用粗天平称量,而配制盐酸溶液要准确标定。为了更加直观地说明以上问题,表 2-6-2 列出温度与盐酸浓度对蔗糖水解速率常数的影响。因为蔗糖及其转化产物都具有旋光性,而且它们的旋光能力不同,故可以利用体系在反应进程中旋光度的变化来度量反应进程。

通过以上讨论可以看出,反应物浓度、温度及催化剂浓度等都对反应速率有影响,但是能影响化学反应速率常数的因素主要是温度、溶剂及催化剂等。

表 2-6-2 温度与盐酸浓度对蔗糖水解速率常数的影响

HCl/(mol·L^{-3})	$k \times 10^3$/min^{-1}		
	25 ℃	35 ℃	45 ℃
0.050 2	0.416 9	1.738	6.213
0.251 2	2.255 0	9.355	35.850
0.413 7	4.043 0	17.000	60.620
0.900 0	11.160 0	46.760	148.800
1.214 0	17.455 0	75.970	

$E = 108$ kJ·mol^{-1}

2. 影响化学反应速率常数的因素

化学反应速率是反应组分的浓度随时间的变化率，$r = \dfrac{dC_B}{v_B dt}$。大多数化学反应的反应速率与反应物浓度的关系方程式为 $r = kC_A^\alpha C_A^\beta \cdots$，其中的比例系数 k 称为化学反应速率常数，它的物理意义是，数值上等于参加反应的物质都处于单位浓度时的反应速率。速率常数 k 的大小与反应物的浓度无关，与反应温度、反应介质、催化剂、反应环境等因素有关。

(1) 温度对反应速率常数的影响

温度对反应速率常数的影响是比较复杂的，目前实验观察到的有五种类型，如图 2-6-1 所示。

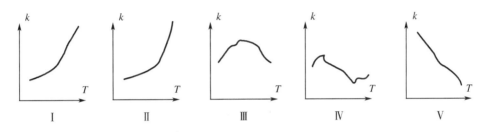

图 2-6-1 温度对反应速率常数影响的几种类型

第 I 种类型，反应速率常数随温度升高而增大，它们之间成指数关系，这类反应最为常见。第 II 种类型，具有爆炸极限的反应，开始时，温度升高，反应速率常数增大，当达到一定温度极限时，反应速率常数增加极快，发生爆炸现象。第 III 种类型常见的是催化反应，如异相催化反应与酶催化反应，由于催化剂有一定的活性温度范围，如 Fe 催化剂活性温度为 500~550 ℃。开始温度升高，催化剂活性增大，反应速率常数增大，达到一定温度范围，催化剂活性达最大值，速率常数最大，超过这个温度范围，催化剂活性下降，反应速率常数也减小。第 IV 种类型是在碳的氢化反应中观察到的，当温度升高时，发生一些副反应而使反应复杂化，反应速率常数变化有起伏。第 V 种类型是比较反常的，温度升高，反应速率常数减少，如 $2NO + O_2 \Longrightarrow 2NO_2$ 反应就是这样。这五种类型的反应中，以第 I 种类反应最为普遍，下面讨论的主要是这种类型的反应。

获得诺贝尔化学奖的范特荷夫 (Van't Hoff) 首先研究了温度对速率常数的影响，得出一个规律，在室温附近，温度每升高 10 ℃，速率常数 k 增加 2~4 倍。Arrhenius 研究了许多反应的速率常数 k 与温度 T 的关系，提出经验方程式 $\ln k = \dfrac{-E_a}{RT} + B$，写成指数式为 $k = Ae^{\frac{-E_a}{RT}}$，其中 A 和 B 是与分子碰撞频率有关的常数，称为指前因子；E_a 是与能量有关的常数，叫作活化能。Arrhenius 方程认为 A 与 E_a 是与温度无关的常数，这是粗略近似的，其实 A 与 E_a 均与温度 T 有关。把上面方程对 T 求微商，可得到微分式 $\dfrac{d\ln k}{dT} = \dfrac{E_a}{RT^2}$，绝大多化学反应的活化能是大于零的，即 $E_a > 0$，因此随着温度的升高，反应速率常数增大。

(2) 溶剂对反应速率常数的影响

在溶液中，反应物分子处于溶剂分子的包围之中，反应物分子必须通过扩散穿过溶剂分子包围圈,与其他反应分子接近、碰撞,才会发生反应,得到产物,而产物又要通过扩散突破周围溶剂分子包围圈离开,避免重新变成反应物分子。在液相反应研究中,把反应物分子在溶剂分子形成的笼子(包围圈)中与溶剂分子的多次碰撞才能从一个笼子中扩散到另一个笼子中的现象称为笼效应,而把两个反应物分子通过扩散移动到同一个笼子中发生连续多次碰撞的现象称为一次偶遇,会形成一个偶遇对。反应物分子扩散冲破溶剂笼子,所需要活化能比较低,不会超过 20 kJ·mol^{-1},而分子之间反应的活化能一般在 40~400 kJ·mol^{-1}。对于活化能较大的反应,扩散对反应速率常数影响较小;而对于活化能较小的反应,扩散对反应速率常数就有较大的影响。溶剂对反应速率常数的影响,体现在扩散对反应速率常数的影响上。而扩散的快慢体现了溶剂分子与反应物分子及产物分子之间作用力的大小。

溶剂对反应速率常数的影响主要体现在以下几方面。

①溶剂的介电常数对速率常数的影响。

溶剂介电常数大,拉开溶质分子正负电荷中心的能力就大,有利于分子解离为正、负离子的反应,而不利于正、负异号离子之间的化合反应。

②溶剂分子的极性对速率常数的影响。

溶剂分子的极性大,与参加反应的离子、极性分子之间有着较强的相互吸引作用,而与参加反应的非极性分子、极性小分子之间作用力较弱,影响反应的速率常数。

③溶剂化作用对速率常数的影响。

一般来说,溶质分子与溶剂分子之间相互作用,或多或少都能形成溶剂化物,属于一种配合物,这就是溶剂化作用。溶剂化作用可以改变反应的活化能,从而影响反应的速率常数。若反应物 A、B 与溶剂产生较强溶剂化作用,形成溶剂化物,降低能量,而偶遇对 A:B 与溶剂是较弱的作用,不形成溶剂化物,那么就是反应的活化能增加,反应速率常数减小,不利于该反应进行。若反应物 A、B 与溶剂之间作用力弱,不形成溶剂物,而偶遇对 A:B 与溶剂之间作用力强,形成稳定溶剂化物,那么降低反应活化能,速率常数增大,有利于该反应进行。

④溶液中离子强度对反应速率常数的影响。

对于溶液电解质之间的反应,即离子之间的反应,不参加反应的其他电解质浓度通过离子强度(I)影响其反应的速率常数。

(3) 催化剂对速率常数的影响

催化剂主要是通过改变反应途径,降低活化能,改变指前因子来影响速率常数的,从而达到加快反应速率的目的。催化反应分为均相催化反应与异相催化反应。均相催化反应的特点是催化剂与反应物混合处于同一相。异相催化反应的特点是催化剂与反应物不处于同一相,如用 Fe 催化合成 NH_3 反应,反应只在催化剂表面附近进行,因此催化剂的表面状态、表面性质对反应速率常数影响很大。

(4) 光对速率常数的影响

在光作用下才能进行的反应称为光化学反应。光化学反应与热化学反应的不同点有二:一是在恒温恒压下,体系的自由能可以增加;二是光化学反应的温度系数很小,即温度对光化学反应的速率常数影响很小,而光的频率、光的强度、光敏剂等对光化学反应的速率常数影响较大。依据光化学基本定律,只有被吸收的光才对化学反应有效;在反应的初级阶段,一个分子吸收一个光子被激活。光化学反应的机理可表示如下

$$A + h\upsilon \rightarrow A^* \text{(初级阶段)}$$

$$A + A^* \rightarrow P \text{（次级阶段）}$$

初级阶段的反应速率与反应物浓度无关,与光的频率、强度有关。

以上主要分析讨论了温度、溶剂、催化剂及光对反应速率常数的影响,除了这些因素外,还有诸如电位等因素对反应速率常数也是有影响的,在这里就不做介绍了。

Experiment 6 Determination of Rate Constant for Hydrolysis of Sucrose

【Purpose of the Experiment】

1. To determine the rate constant and half-life of hydrolysis of sucrose.
2. To know the effects of the catalyst for reaction rate and rate constant.

【Principle】

The reaction equation for hydrolysis of sucrose is:

$$C_{12}H_{22}O_{11} + H_2O \xrightarrow{H^+} C_6H_{12}O_6 + C_6H_{12}O_6 \quad (2-6-1)$$
$$\text{Sucrose} \qquad\qquad\qquad \text{Glucose} \quad \text{Fructose}$$

In pure water, the reaction rate for hydrolysis of sucrose is extremely slow. In order to accelerate the hydrolysis reaction, the reaction is often carried out in an acidic medium. As a second-order reaction, the reaction rate is proportional to one power of the sucrose concentration and one power of the water concentration. However, the water as a reactant is still present in a large amount compared to a low sucrose concentration, and its concentration changes can be ignored. Therefore, the hydrolysis of sucrose can be regarded as a first-order reaction (specifically, a "quasi-first-order reaction").

The first-order reaction rate equation is:

$$-\frac{dC}{dt} = kC \quad (2-6-2)$$

Integral type:

$$t = \frac{1}{k}\ln\frac{C_0}{C} \quad (2-6-3)$$

Half life:

$$t_{1/2} = \frac{\ln 2}{k} = \frac{0.6931}{k} \quad (2-6-4)$$

In order to obtain the concentration of sucrose during the reaction, the physical parameter of optical rotation is employed. In this experiment, sucrose, glucose and fructose have different optical rotation values. Therefore, the change of optical rotation in the reaction process can be used to measure the progress of the reaction. The instrument used to measure the optical rotation of a substance is called a polarimeter. The optical rotation of the solution is related to the optical rotation ability, solvent property, solution concentration, sample length and temperature of the optically active substance contained in the solution. When other conditions are fixed, the optical rotation is linear with the reactant concentration (C), ie:

$$\alpha = BC \qquad (2-6-5)$$

Where the proportionality constant B is related to the optical rotation ability of the substance, the solvent property, the wavelength of the light source, the length of the sample tube, the temperature at the time of the reaction.

The optical rotation of a substance is the specific optical rotation, which can be expressed by the Equation (2-6-6).

$$[\alpha]_D^{20} = \frac{100\alpha}{lC} \qquad (2-6-6)$$

Where 20 in the upper right corner indicates that the measured temperature is 20 ℃. D means that the light source used is the D line of the sodium light source. α is the measured optical rotation, (°). l is the length of the sample tube, dm. C is the concentration of sucrose, g/100 mL.

The sucrose as a reactant is a dextrorotatory substance, which has a specific optical rotation of $[\alpha]_D^{20} = 66.6°$. Glucose is also a dextrorotatory substance with the specific optical rotation of $[\alpha]_D^{20} = 52.5°$. But fructose is a levorotatory substance with the specific optical rotation of $[\alpha]_D^{20} = -91.9°$. Therefore, the specific optical rotation of the reaction solution gradually decrease to a negative value with the increase of the reaction time.

So, specific optical rotation at $t=0, t=\infty$ and $t=t$ can be expressed as:

$$\alpha_0 = A_R C_0 \qquad (2-6-7)$$
$$\alpha_\infty = A_P C_0 \qquad (2-6-8)$$
$$\alpha_t = A_R c + A_P (C_0 - C) \qquad (2-6-9)$$

Then,

$$C = A(\alpha_t - \alpha_\infty) \qquad (2-6-10)$$

Where A_R, A_P and A are the proportionality constants.

Bringing the concentration c into the Equation (2-6-3) is given the following equation:

$$\lg(\alpha_t - \alpha_\infty) = -\frac{k}{2.303}t + \lg(\alpha_0 - \alpha_\infty) \qquad (2-6-11)$$

Where α_0, α_t and α_∞ are the specific optical rotations of time $t=0, t$ and ∞, respectively.

It can be seen from the Equation (2-6-11) that if a straight line is obtained by plotting t, the rate constant k of the reaction can be obtained from the slope of the fitted straight line for $\lg(\alpha_t - \alpha_\infty) - t$.

【Apparatus and Reagents】

Automatic polarimeter (WZZ-2B); Electric balance; Stopwatch; 25 mL of pipette; 20 cm of sample tube; 50 mL of Volumetric flask; Sucrose; 2.0 mol·L^{-1} HCl solution.

【Procedures】

1. Turn on the power to the automatic polarimeter and warm up for 10 min. Wash and fill the

sample tube with distilled water, respectively. The sample tube containing distilled water is then placed in the polarimeter to calibrate the zero point.

2. Rotation determination for hydrolysis of sucrose (α_t)

Pipette 25 mL of 20% sucrose solution and 25 mL of 2.0 mol·L^{-1} HCl solution into two clean and dry conical flasks, respectively. The two conical flasks are placed in the water bath at 25℃ for 10 min. Rapidly add 25 mL of 20% sucrose solution into 25 mL of 2.0 mol·L^{-1} HCl solution, shake the mixed solution and start the stopwatch at the same time. Subsequently, rinse the sample tube twice with 1~2 mL of the mixed solution. The sample tube is filled with the mixed solution and then placed in the the polarimeter to measure optical rotation. Record the optical rotation value (α_t) of 5, 10, 15, 20, 30, 50, 75 and 100 min and fill in Table 2-6-1.

3. Determination of α_∞

To completely hydrolyze sucrose, the mixed solution can be placed in a 50~60 ℃ water bath and heated for 40 min to accelerate the hydrolysis of sucrose, and then cooled to the measured temperature. A α_∞ value can be obtained by measuring the cooled solution.

【Records of Data】

1. Record the experimental data in Table 2-6-1.

Room temperature: _____ ; Atmospheric pressure: _____ ;
Reaction temperature: _____ ;   Concentration of HCl: _____ ;
Length of the sample: $l =$ _____ ; α_∞ : _____ .

Table 2-6-1 Experimental data of hydrolysis of sucrose

Reaction time/min	α_t	$\alpha_t - \alpha_\infty$	$\lg(\alpha_t - \alpha_\infty)$	Rate constant (k)

【Data Analysis】

1. Plot the $\lg(\alpha_t - \alpha_\infty) - t$ curve. Calculation the reaction rate constants (k at 25 ℃) from the slopes of the plotted curves.

2. Calculate the half life of hydrolysis of sucrose.

【Notes】

1. The temperature of the water bath in step 3 should not be too high, otherwise a side reaction will occur, the color of the solution will turn yellow, and the stopper should be covered during the heating process to prevent the solution from evaporating to affect the concentration.

2. At the end of the experiment, the sample tube should be washed and dried immediately to prevent acid corrosion of the sample tube.

【Post Lab Questions】

1. What are the factors involved in the rate for hydrolysis of sucrose?

2. Is the concentration of sucrose solution not accurate enough to affect the measurement results? Why?

实验7 原电池电动势的测定

【实验目的】

1. 掌握电解法制备 Cu、Zn 电极；
2. 了解可逆电池、可逆电极、盐桥等概念；
3. 掌握利用能斯特方程计算电池的理论电动势；
4. 掌握对消法测量电动势的原理及电位差计测定电池电动势的方法。

【预习要求】

1. 了解电解法制备 Cu、Zn 电极的原理及方法；
2. 明确可逆电池、可逆电极的概念；
3. 了解不同盐桥的使用条件；
4. 能够利用能斯特方程计算电池的理论电动势；
5. 了解电位差计的测量原理及使用方法。

【实验原理】

电池由正、负两极组成，电池放电过程中，正极发生还原反应，负极发生氧化反应，电池内部则发生其他反应（如发生离子迁移），电池反应是电池所有反应的总和。

电池除了可以用来作为电源对外做电功，还可以用来研究构成此电池的化学反应的热

力学性质。由化学热力学可知,在恒温、恒压、可逆条件下,电池反应的吉布斯函数改变值为

$$\Delta G_{T,p} = -nEF \quad (2-7-1)$$

式中,n 为进行单位反应时,电池中电极上得失电子的物质的量;E 是可逆电池的电动势;F 是法拉第常数。

所以若能测定该电池的电动势 E,则可求出电池反应的吉布斯函数变化 ΔG,进而求出 ΔH、ΔS 等其他热力学函数。但必须注意,只有在恒温、恒压、可逆条件下,式(2-7-1)才能成立。这就要求电池符合可逆电池的三个条件:①电池反应本身是可逆的,即化学可逆性;②电路里电流趋于零,即热力学可逆;③不存在任何不可逆的液接界,即实际可逆性。

实际情况中,原电池中往往存在液接界,精度要求不高的情况下常用"盐桥"来减小其带来的不可逆性。所谓"盐桥"是指一种正负离子迁移数比较接近的盐类溶液所构成的桥,用来连接原来产生显著接界电势的两个液体,从而使其彼此不直接接界。使用盐桥后,液接界电位一般在毫伏数量级,常用的盐桥有 KCl、KNO_3、NH_4NO_3 等。

在进行电池电动势测量时,电池反应在接近热力学可逆条件下(电流 $I \to 0$),可用电位差计来测量。电位差计测定电池电动势的原理是:电位差计提供一个与待测电池大小相等、方向相反的电动势,从而抵消待测电池的电动势,进而使电路中的电流趋于零,此时电位差计的示数即为被测电池的电动势(图 2-7-1)。

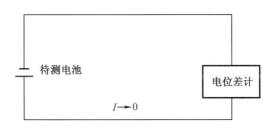

图 2-7-1 电动势测量原理示意图

测量电池电动势时不能盲目调节电位差计,而是应该根据被测电池的理论电动势大小在其附近进行调节,否则可能远远超出量程。所以,首先需要对电池的理论电动势进行计算。方法一,对电池反应整体应用能斯特方程;方法二,先用能斯特方程分别计算正负电极的电势 φ_+、φ_-,然后由 $E = \varphi_+ + \varphi_-$ 求得。

以 Cu-Zn 电池为例:

电池结构: $Zn(s) | ZnSO_4(\alpha_{Zn^{2+}}) \| CuSO_4(\alpha_{Cu^{2+}}) | Cu(s)$

负极反应: $Zn(s) \longrightarrow Zn^{2+}(\alpha_{Zn^{2+}}) + 2e^-$

正极反应: $Cu^{2+}(\alpha_{Cu^{2+}}) + 2e^- \longrightarrow Cu(s)$

总电池反应: $Zn(s) + Cu^{2+}(\alpha_{Cu^{2+}}) \longrightarrow Zn^{2+}(\alpha_{Zn^{2+}}) + Cu(s)$

方法一:由能斯特方程,该电池的电动势为

$$E = E^{\ominus} - \frac{RT}{2F} \ln \frac{\alpha_{Zn^{2+}} \cdot \alpha_{Cu}}{\alpha_{Cu^{2+}} \cdot \alpha_{Zn}} \quad (2-7-2)$$

纯固体的活度等于1,即

$$\alpha_{Zn} = \alpha_{Cu} = 1 \qquad (2-7-3)$$

代入式(2-7-2)得

$$E = E^{\ominus} - \frac{RT}{2F}\ln\frac{\alpha_{Zn^{2+}}}{\alpha_{Cu^{2+}}} \qquad (2-7-4)$$

式中,$E^{\ominus}$ 为溶液中锌离子的活度 $\alpha_{Zn^{2+}}$ 和铜离子的活度 $\alpha_{Cu^{2+}}$ 均等于1时的电池电动势。

方法二:先根据能斯特方程求出电极电势。

Zn 电极电势:
$$\varphi_{-} = \varphi^{\ominus}_{Zn^{2+},Zn} - \frac{RT}{2F}\ln\frac{\alpha_{Zn}}{\alpha_{Zn^{2+}}} \qquad (2-7-5)$$

Cu 电极电势:
$$\varphi_{+} = \varphi^{\ominus}_{Cu^{2+},Cu} - \frac{RT}{2F}\ln\frac{\alpha_{Cu}}{\alpha_{Cu^{2+}}} \qquad (2-7-6)$$

式中,$\varphi^{\ominus}_{Cu^{2+},Cu}$、$\varphi^{\ominus}_{Zn^{2+},Zn}$ 是当 $\alpha_{Zn^{2+}} = \alpha_{Cu^{2+}} = 1$ 时,Cu 电极和 Zn 电极的标准电极电势(参见附录)。然后再根据

$$E = \varphi_{+} - \varphi_{-} \qquad (2-7-7)$$

求出电池的电动势。注意:由于电极电势定义为还原电势,即都发生还原反应,所以电极的能斯特方程的活度项中都为还原产物的活度比氧化产物的活度,以 Cu 为例即 $\alpha_{Cu}/\alpha_{Cu^{2+}}$。

在电化学中,电极电势的绝对值至今还无法测定,而是以某一电极的电极电势作为零标准,然后将待测的电极与其组成电池,测定此电池的电动势,则该电动势即为该被测电极的电极电势。被测电极在电池中的正负极性,可由它在零标准电极两者的还原电位比较而定。通常将氢电极在氢气压力为1大气压,溶液中 $\alpha_{H^{+}}$ 为1时的电极电位规定为零(称为标准氢电极),然后与其他被测电极进行比较。

由于使用氢电极不方便,一般常取另外一些制备工艺简单、易于复制、电位稳定的电极作为参比电极来代替氢电极,常用的有甘汞电极、氯化银电极等。这些电极与标准氢电极比较而得到的电位已精确测出,见相关阅读部分。

根据 $\Delta G_{T,p} = -nEF$,如果测定了电池的电动势 E 即可得到电池反应的吉布斯函数变化。此外,根据

$$\Delta_{r}S_{m} = -\left(\frac{\partial \Delta_{r}G_{m}}{\partial T}\right)_{p} = nF\left(\frac{\partial E}{\partial T}\right)_{p} \qquad (2-7-8)$$

测定不同温度下的电动势 E,作线性拟合则可得到电动势随温度的变化率 $\partial E/\partial T$,从而可以计算电池反应的熵变。进一步,根据

$$\Delta_{r}H_{m} = \Delta_{r}G_{m} + nFT\left(\frac{\partial E}{\partial T}\right)_{p} \qquad (2-7-9)$$

可以计算得到电池反应的焓变。所以,通过测量电池的电动势可以得到电池反应的热力学函数变化,这也是测定化学反应热力学的一个重要手段。

本实验中,我们要测定的是 Cu-Zn、Cu-甘汞、Zn-甘汞三组电池的电动势,其中所使用的甘汞电极为饱和甘汞电极,其结构及特性在理论课教材中进行了介绍,在本实验的相关阅读中也有更为详尽的介绍。而 Zn 电极及 Cu 电极则需要在实验中通过电解法制备。

电解是在外加电压下,电流通过电解质溶液(或熔融电解质),在阳极、阴极上发生氧化-还原反应的过程,在本实验中,我们将通过电解 $ZnSO_4$ 和 $CuSO_4$ 溶液,还原 Zn^{2+} 和 Cu^{2+} 使 Zn

和 Cu 沉积在石墨棒表面,进一步将其作为电极组成原电池,并测定电动势。

【仪器与试剂】

SDC-Ⅱ 数字电位差综合测试仪 1 台,直流电源 1 台,超级恒温水浴 1 套,标准电池 1 块,饱和甘汞电极,250 mL 三口瓶 1 个,50 mL 烧杯 3 支,$CuSO_4 \cdot 5H_2O$,$ZnSO_4 \cdot 7H_2O$,NaOH,石墨棒。

【实验步骤】

1. 电极的制备

电解法制备 Cu、Zn 电极的装置如图 2-7-2 所示。

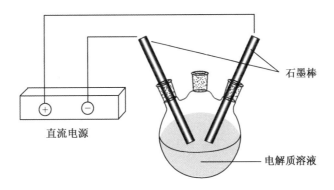

图 2-7-2 电解法制备 Cu、Zn 电极的装置

(1) Cu 电极的制备:取 15 cm 左右石墨棒 2 根,用软布或纸擦拭石墨棒表面及两端,250 mL 三口瓶中加入 100 mL 的蒸馏水,然后加入过量的 $CuSO_4 \cdot 5H_2O$ 配制成饱和溶液,将石墨棒与直流电源以带鳄鱼夹的导线连接,并分别置于三口瓶的两个口中(注意:两个石墨棒不能接触,以防短路)。将直流电源的电流控制旋钮调至最大,电压旋钮调至最小,启动电源,调节电压至 4 V(过大的电压可能导致析出的铜较为松散),将有 Cu 在电极上析出,持续 15 min 左右,得到镀有 Cu 的石墨棒,即 Cu 电极。

(2) Zn 电极的制备:取 11.5 g $ZnSO_4 \cdot 7H_2O$ 加入 100 mL 蒸馏水,充分溶解后,加入到 100 mL 三口烧瓶中,取 15 cm 左右石墨棒 2 根,用软布或纸擦拭石墨棒表面及两端,将石墨棒与直流电源以带鳄鱼夹的导线连接,并分别置于三口瓶的两个口中(注意:两个石墨棒不能接触,以防短路),将直流电源的电流控制旋钮调至最大,电压旋钮调至最小,启动电源,调节电压至 4 V,电解 20 min 左右,得镀有 Zn 的石墨棒,即 Zn 电极。

(3) 饱和甘汞电极:观察甘汞电极里是否有 KCl 晶体,如果没有,则需要从侧管口加入少量 KCl 固体,确保饱和 KCl 溶液能够浸没电极上端的汞,并且电极内没有气泡存在。

(4) 电池的组合:分别量取约 20 mL 浓度为 0.1 mol/L 的 $CuSO_4$ 及 0.1 mol/L 的 $ZnSO_4$ 溶液置于小烧杯中,以琼脂-饱和 KCl 盐桥连接两个烧杯,再将上面制备的 Cu 电极和 Zn 电极分别置于盛有 $CuSO_4$ 的小烧杯内,即成 Cu-Zn 电池:

$$Zn(s) \mid ZnSO_4(0.1 \text{ mol} \cdot L^{-1}) \parallel CuSO_4(0.1 \text{ mol} \cdot L^{-1}) \mid Cu(s)$$

同法组成下列电池:

$$Zn(s) | ZnSO_4(0.1\ mol \cdot L^{-1}) \| KCl(饱和) | Hg_2Cl_2(s) | Hg$$
$$Hg | Hg_2Cl_2 | KCl(饱和) \| CuSO_4(0.1\ mol \cdot L^{-1}) | Cu(s)$$

2. 电池电动势的测量

SDC-Ⅱ数字电位差综合测试仪的操控面板如图2-7-3所示。

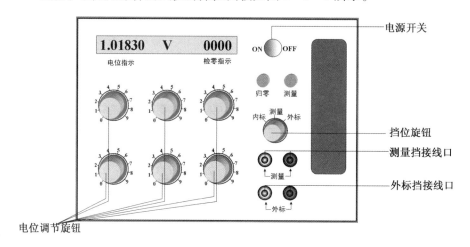

图2-7-3 SDC-Ⅱ数字电位差综合测试仪的操控面板

(1)打开电源开关。
(2)将挡位旋钮旋至外标挡。
(3)用导线将标准电池与电位差计的"外标挡接线口"连接。
(4)调节电位调节旋钮,使输出电位为1.018 30 V(标准电池电动势的温度校正公式: $E_t = 1.018\ 30 - 0.000\ 040\ 6 \times (t - 20)$),观察检零窗口示数,如不为零,则按"归零"按钮将其归零。
(5)将接位旋钮旋至测量挡。
(6)将待测电池与"测量挡接线口"连接(注意电池正负极)。
(7)根据待测电池的理论电动势调节"电位调节旋钮",当输出电位等于待测电池的真实电动势时,检零指示窗口示数为零,此时"电位指示"窗口的示数即为该电池的电动势。
(8)测定不同温度下电池的电动势,温度范围为15~50 ℃,每隔5~10 ℃测一次,电池用恒温水浴恒温。

【实验记录及数据处理】

1. 按式(2-7-5)、式(2-7-6)、式(2-7-7)计算所测电池电动势的理论值,列于表2-7-1中。计算时,物质的浓度应以下式计算:

$$\alpha_{Zn^{2+}} = \gamma_{\pm} c_{Zn^{2+}},\ \alpha_{Cu^{2+}} = \gamma_{\pm} c_{Cu^{2+}} \qquad (2-7-10)$$

其中,$\gamma_{\pm}$是离子的平均离子活度系数,其数值见表2-7-2。

表2-7-1 电动势测定数据结果

电池		$E_{测量值}$/V	$E_{理论值}$/V	相对误差/%
1	Zn(s)｜ZnSO$_4$(0.1 mol·L^{-1})‖CuSO$_4$(0.1 mol·L^{-1})｜Cu(s)			
2	Zn(s)｜ZnSO$_4$(0.1 mol·L^{-1})‖KCl(饱和)｜Hg$_2$Cl$_2$｜Hg			
3	Hg｜Hg$_2$Cl$_2$｜KCl(饱和)‖CuSO$_4$(0.1 mol·L^{-1})｜Cu(s)			

表2-7-2 不同浓度下CuSO$_4$及ZnSO$_4$的活度系数

电解质	浓度	
	0.1 mol·L^{-1}	0.01 mol·L^{-1}
CuSO$_4$	0.16	0.40
ZnSO$_4$	0.15	0.387

2.将所得电池电动势与绝对温度作图,并由图上的曲线求取18 ℃、25 ℃、35 ℃三个温度下的E和$(dE/dT)_p$,$(dE/dT)_p$即为曲线切线的斜率。

3.利用式(2-7-1)、式(2-7-8)、式(2-7-9)计算18 ℃、25 ℃、35 ℃时的$\Delta_r G_m$、$\Delta_r H_m$、$\Delta_r S_m$,并与文献值相比较。

【讨论与说明】

电动势的测量方法属于平衡测量,在测量过程中尽可能地做到在可逆条件下进行。为此应注意以下几点:

1.测量需要计算电池的理论电动势,以便在测量时能迅速找到平衡点,这样可以避免电极极化。

2.为判断所测量的电动势是否为平衡电势,一般应在15 min左右的时间内,等间隔地测量7~8个数据。若这些数据是在平均值附近摆动,偏差小于±0.000 5 V,则可以认为已达平衡,可取其平均值作为该电池的电动势。

3.连接线路时注意正负极和盐桥的正确使用。

【思考题】

1.测定电动势为何使用电位差计? 使用万用表是否可以?
2.测定电动势为何使用"盐桥"?
3.对消法测量电池电动势的主要原理是什么?

【相关阅读】

1.液接界电势

许多实用的电池中两个电极周围的电解质溶液的性质不同(如参比电极内的溶液和被研究电极内溶液的组成不一样,或者两种溶液相同而浓度不同等),它们不处于平衡状态。

当这两种溶液相接触时,存在一个液体接界面,在接界面的两侧,会有离子往正反方向扩散,随着时间的延长,最后扩散达到相对稳定。这时,在液接面上产生一个微小的电势差,这个电势差称为液接界电势。

如两种不同浓度的 HCl 溶液的界面上,H^+ 和 Cl^- 有浓度梯度的突跃,因此,两种离子必从浓的一边向稀的一边扩散。因为 H^+ 比 Cl^- 的淌度大得多,所以最初 H^+ 以较高的速率进入较稀的一相。这个过程使稀相出现过剩的 H^+ 而带正电荷;而浓相有过剩的 Cl^- 而带负电荷,结果产生了界面电势差。由于电势差的存在,该电场使 H^+ 的扩散速率减慢,同时加快了 Cl^- 的扩散速率,最后这两种离子的扩散速率相等,此时在界面上得到一个微小的稳态电势,即液接界电势。根据它产生的原因,有时也称为扩散电势。

液接界电势至今无法精确测量和计算,但在稀溶液中,使用亨德森(Henderson)公式可以满足一般的要求:

$$E_J = \frac{(u_1 - v_1) - (u_2 - v_2)}{(u_1' + v_1') - (u_2' + v_2')} \cdot \frac{RT}{F} \ln \frac{u_1' + v_1'}{u_2' + v_2'}$$

式中 $\begin{cases} u = \sum m_+ \lambda_{m+} \\ u' = \sum m_+ \lambda_{m+} Z_+ \end{cases}$ $\begin{cases} v = \sum m_- \lambda_{m-} \\ v' = \sum m_- \lambda_{m-} Z_- \end{cases}$

m_+ 和 m_- 分别为阳离子和阴离子的质量摩尔浓度;λ_{m+} 和 λ_{m-} 分别为阳离子和阴离子的摩尔电导率;Z_+ 和 Z_- 分别为阳离子和阴离子的价数;注脚"1"和"2"表示溶液 1 和 2;E_J 为液接界电势。

在水溶液中,两种不同溶液的 E_J 一般小于 50 mV,如 1 mol·L^{-1} NaOH 与 0.1 mol·L^{-1} KCl 两溶液间的 E_J 值,按亨德森公式计算,$E_J = 45$ mV。可见液接界电位是不可被忽视的。在测量电极电势时,要采取措施减小液接界电势。减小液接界电势的方法一般采用盐桥。

2. 盐 桥

在电池电动势测量中,为减少不同电解质溶液(不同种电解质或同种电解质浓度不同)接界处的液体接界电势差(或称扩散电势差),常以盐桥跨接,将两种溶液隔开,如图 2-7-4 所示。盐桥管内是电解质的浓溶液,常用琼脂为骨架材料。饱和 KCl、NH_4NO_3 或 KNO_3 溶液,这些电解质溶液的正、负离子迁移数很接近,与此浓溶液相比,溶液 1 和 2 的浓度相对很低。在盐桥与溶液 1 和 2 接界处,主要的扩散是盐桥中正、负离子进入溶液 1 和 2 中,因其正、负离子迁移数很相近,在两个液体接界处产生的液接电势差大小相近,但符号相反,可以有所抵消,使总的液体接界电势差降低。

盐桥中的浓电解质溶液,不仅要求其正、负离子的迁移数很接近,而且其正、负离子不能与所使用体系的溶液起反应,选用何种电解质的盐桥,视具体电池而定。如用 KCl 作电解质的盐桥,不能用于含有 Ag^+、Hg_2^{2+}、ClO_4^-、K^+ 的溶液。

盐桥的制作及使用:以琼脂-饱和 KCl 盐桥为例。将 3 g 琼脂溶于 97 g 去离子水中,慢慢在水浴中加热,至完全溶解,加入 30 g KCl,充分搅拌,当 KCl 完全溶解后,趁热用滴管或虹吸管将此溶液装入已洗净的 U 形玻璃管中,注意避免管内有气泡,并使管口平整。静置,待冷却后琼脂呈凝固态便成为盐桥,浸在饱和 KCl 溶液中备用。

高浓度的酸、氨都会与琼脂作用,破坏盐桥、沾污溶液,此种情况不能使用琼脂盐桥。为防止盐桥溶液中的离子对待测体系的影响,可采用多道盐桥,适合待测体系的盐桥与减少接界电势差的盐桥同时使用,如图 2-7-4 所示。

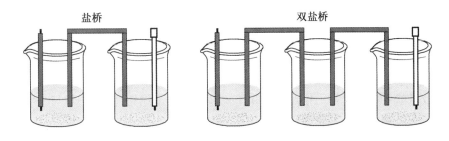

图 2-7-4 盐桥的使用

3. 参 比 电 极

电极电势的测量是通过被测电极与参比电极组成电池测其电池的电动势,然后根据参比电极的电势求得被测电极的电极电势。电极电势的测量除了要考虑电动势测量中的有关问题之外,还要特别注意参比电极的选择。

（1）参比电极的选择

选择参比电极必须注意下列问题：

①参比电极必须是可逆电极,它是电极电势也是可逆电势。

②参比电极必须具有良好的稳定性和重现性,即它的电极电势受放置时间(一般为数天)影响不大,各次制作的同样的参比电极,其电极电势也应基本相同。

③由金属和金属难溶盐或金属氧化物组成的参比电极属第二电极,如银-氯化银电极、汞-氧化汞电极,要求这类金属的盐或氧化物在溶液中的溶解度很小。

④参比电极的选择必须根据被测体系的性质来决定。例如,氯化物体系可选甘汞电极或氯化银电极,硫酸溶液体系可选硫酸亚汞电极,碱性溶液体系可选氧化汞电极等。在具体选择时还必须考虑减小液接界电势等问题。此外还可以采用氢电极作参比电极。

（2）水溶液体系常用的参比电极

①氢电极

氢电极主要用作电极电势的标准。但在酸性溶液中也可作为参比电极,尤其在测量氢超电势时,采用同一溶液中的氢电极作为参比电极,可简化计算。

氢电极的电极反应：

在酸性溶液中：

$$2H^+ + 2e^- == H_2(g)$$

在碱性溶液中：

$$2H_2O + 2e^- == H_2(g) + 2OH^-$$

氢电极的电极电势与溶液的 pH 值和氢气压力有关。

$$\varphi_{H^+/H_2} = \frac{RT}{F} \ln \frac{a(H^+)}{p_{H_2}^{1/2}}$$

式中,$a(H^+)$ 为 H^+ 离子的活度；p_{H_2} 为氢气的压力(p_{H_2} = 大气压 - 水的饱和蒸气压),如果氢

气的压力是 101.325 kPa(即标准大气压力),在 25 ℃时氢电极的电极电势为

$$\varphi_{H^+/H_2} = -0.05916 \text{ pH}$$

氢电极的优点是其电极电势仅决定于液相的热力学性质,因而易做到实验条件的重复。但其电极反应在许多金属上的可逆程度很低,因此必须选择对此反应有催化作用的惰性金属作为电极材料。一般采用大小适中(如$(1 \times 1)\text{cm}^2$)的金属铂片,将铂片与一铂丝相焊接,铂丝的另一头可烧结在 5 号量器玻璃管中。

②甘汞电极

标准氢电极虽然稳定,但操作麻烦,所以常用重现性好又比较稳定的甘汞电极作为参比电极。甘汞电极的组成为

$$\text{Hg} \mid \text{Hg}_2\text{Cl}_2 \mid \text{KCl}(\text{溶液})$$

其电极反应为

$$\text{Hg}_2\text{Cl}_2 + 2e^- \Longrightarrow 2\text{Hg} + 2\text{Cl}^-$$

甘汞电极是汞和甘汞与不同浓度的 KCl 溶液组成的电极,它的电极电势可以与标准氢电极组成电池而精确测定,所以又称这种电极为二级标准电极。它的电极电势随 Cl^- 的活度不同而不同,通常使用的有 $0.1 \text{ mol} \cdot \text{L}^{-1}$、$1.0 \text{ mol} \cdot \text{L}^{-1}$ 和饱和式三种,其电极电势可参考本书附录。KCl 溶液浓度为 $0.1 \text{ mol} \cdot \text{L}^{-1}$ 的甘汞电极的电动势温度系数小,但饱和氯化钾的甘汞电极容易制备,而且使用时可以起盐桥的作用,所以平时用得较多。

甘汞电极的结构形式有多种,图 2-7-5 列出市售的(a)(b)两种和实验室制作的(c)(d)两种。

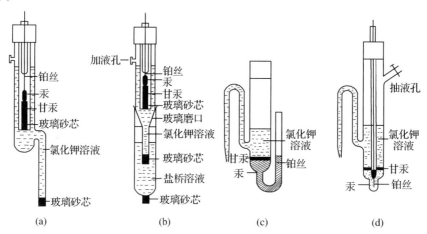

图 2-7-5 甘汞电极的几种形式

③银-氯化银电极

银-氯化银电极为 $\quad\quad\quad\quad\quad \text{Ag} \mid \text{AgCl} \mid \text{Cl}^-(\text{溶液})$

其电极反应为

$$AgCl + e^- =\!=\!= Ag + Cl^-$$

其电极电位取决于 Cl^- 的活度。该电极具有良好的稳定性和较高的重现性,且无毒、耐震。其缺点是必须浸于溶液中,否则 AgCl 层会因干燥而剥落。另外,AgCl 遇光会分解,所以银-氯化银电极不易保存。其电极电势见表2-7-3。

表2-7-3 银-氯化银电极的电极电势

电极	温度/℃	电极电势/V
$Ag\mid AgCl\mid Cl^-[a(Cl^-)=1.0\ mol\cdot L^{-1}]$	25	0.222 3
$Ag\mid AgCl\mid KCl\ (0.1\ mol\cdot L^{-1})$	25	0.288 0
$Ag\mid AgCl\mid KCl$(饱和)	25	0.198 1
$Ag\mid AgCl\mid KCl$(饱和)	60	0.165 7

④汞-硫酸亚汞电极

汞-硫酸亚汞电极由汞、硫酸亚汞和含有 SO_4^{2-} 离子的溶液组成,即

$$Hg\mid Hg SO_4\mid SO_4^{2-}(溶液)$$

其电极反应为

$$Hg_2SO_4 + 2e^- =\!=\!= SO_4^{2-} + 2Hg$$

汞-硫酸亚汞电极常用作硫酸体系的参比电极,如铅蓄电池的研究、硫酸介质中的金属腐蚀的研究等。

⑤汞-氧化汞电极

汞-氧化汞电极是碱性溶液中常用的参比电极,由汞、氧化汞和碱性溶液组成,即

$$Hg\mid HgO\mid OH^-(溶液)$$

其电极反应为

$$HgO + H_2O + 2e^- =\!=\!= Hg + 2OH^-$$

其电极结构和形式与甘汞电极基本相同。由于在碱性溶液中一价汞离子会被歧化为零价汞和二价汞离子,所以体系中不会因 Hg_2O 的存在而引起电势的偏移。它是一个重现性很好的电极。

4. 标 准 电 池

标准电池的电动势具有很好的重现性和稳定性,其重现性一般能达到 0.1 mV。稳定性是指两种情况,一是当电位差计电路内有微量不平衡电流通过该电池时,由于电池的可逆性好,电极电势不发生变化,电池电动势仍能保持恒定;二是能在恒温条件下较长时期内保持电动势基本不变。

标准电池可分饱和式和不饱和式两类,前者可逆性好,因而电动势的重现性、稳定性均好,但温度系数较大,必须进行温度校正,一般用于精密测量中;后者的温度系数很小,但可逆性差,用在精度要求不高的测量中,可以免除烦琐的温度校正。

饱和式标准电池的电化学式表示为

镉汞齐$\{w(\text{Cd})=0.125\}$ | $\text{CdSO}_4 \cdot \dfrac{8}{3}\text{H}_2\text{O}(s)$ | CdSO_4 饱和溶液 | $\text{Hg}_2\text{SO}_4(s)$ | Hg

其电极反应为

$$阳极:\text{Cd}(\text{Hg}) = \text{Cd}^{2+} + \text{Hg}(l) + 2e^-$$
$$阴极:\text{Hg}_2\text{SO}_4(s) + 2e^- = 2\text{Hg}(l) + \text{SO}_4^{2-}$$

电池反应为

$$\text{Hg}_2\text{SO}_4(s) + \text{Cd}(\text{Hg})(a) + \dfrac{8}{3}\text{H}_2\text{O} = \text{CdSO}_4 \cdot \dfrac{8}{3}\text{H}_2\text{O}(s) + 3\text{Hg}(l)$$

标准电池在使用过程中,不可避免地会有充、放电流通过,使电极电势偏离其平衡电势值,导致整体电动势的改变。虽然饱和式标准电池的去极化能力较强,充、放电流结束后电动势的恢复也较快,但仍应对通过标准电池的电流严格限制在允许的范围内。

由于标准电池的温度系数与正负两极都有关系,故放置时必须使两极处于同一温度。饱和标准电池中的 $\text{CdSO}_4 \cdot \dfrac{8}{3}\text{H}_2\text{O}(s)$ 晶粒在温度波动的环境中会反复不断地溶解、再结晶,致使原来很微小的晶粒结成大块,增加了电池的内阻,降低了电位差计中检流计回路的灵敏度。因此,应尽可能将标准电池置于温度波动不大的环境中。机械振动会破坏标准电池的平衡,在使用及搬移时应尽量避免震动,绝对不允许倒置。光会使 Hg_2SO_4 变质,此时标准电池仍可能具有正常的电动势值,但其电动势对于温度变化的滞后特性较大,因此标准电池应避免光照。

5. 电解时的电极反应

由于水溶液中存在 H^+ 和 OH^-,所以电解时,不仅电解质可能发生电极反应,H^+ 和 OH^- 也可能同时发生反应。通常情况下,电解时阳极上总是优先发生极化电势较低的反应,阴极上总是发生极化电势较高的反应。所以,在考虑电极本身的电势的同时,还要考虑超电势的存在。以本实验电解 ZnSO_4 为例,若不考虑超电势,则 Zn 电极的电势约为 -0.7620 V,而 H 电极的电势约为 $0 \sim -0.414$ V(pH $= 0 \sim 7$),H 电极的电势较高,所以 H^+ 应首先在阴极被还原,但由于 H 在 Zn 电极上有较高的超电势,所以根据

$$E_{阴} = E_{阴,平} - \eta_{阴}$$

H 电极的极化电极电势反而更低,所以最终是 Zn^{2+} 在阴极被还原析出。

Experiment 7　Determination of the Electromotive Force of Galvanic Cells

【Purpose of the Experiment】

(1) To master the preparation of Cu and Zn electrodes by electrolysis.

(2) To learn the concepts of reversible batteries, reversible electrodes and salt bridges.

(3) To master the theoretical electromotive force of galvanic cells by using the Nernst

equation.

(4) To master the principle of measuring the electromotive force by the elimination method and the method of measuring the electromotive force of galvanic cells by the potentiometer.

【Principle】

The galvanic cell consisting of two electrodes dipping into an electrolyte solution is widely used to determine activity coefficient, dissociation constant, solubility constant, some thermodynamic functions, and so on. A galvanic cell is a device converting chemical energy to electric energy, in which oxidation occurs on the anode, while reduction takes place on the cathode.

When measuring galvanic cell electromotive force, the potentiometer cannot be blindly adjusted. Instead, it should be adjusted in the vicinity according to the theoretical electromotive force of the battery under test. Otherwise, it may be far beyond the range. Therefore, it is first necessary to calculate the theoretical electromotive force of the battery. Method 1: Apply the Nernst equation to the overall reaction of the battery; Method 2: First calculate the two potentials of the positive and negative electrodes using the Nernst equation, and then find the $E = \varphi_+ - \varphi_-$.

Herein Daniell cell is given as an example to demonstrate the expression of cells as follow:

Battery structure: $Zn(s) | ZnSO_4(\alpha_{Zn^{2+}}) \| CuSO_4(\alpha_{Cu^{2+}}) | Cu(s)$

Negative reaction: $Zn(s) = Zn^{2+}(\alpha_{Zn^{2+}}) + 2e^-$

Positive reaction: $Cu^{2+}(\alpha_{Cu^{2+}}) + 2e^- = Cu(s)$

Total battery reaction: $Zn(s) + Cu^{2+}(\alpha_{Cu^{2+}}) = Zn^{2+}(\alpha_{Zn^{2+}}) + Cu(s)$

Method 1: The Nernst equation, the electromotive force of galvanic cell.

$$E = E^\ominus \frac{RT}{2F} \ln \frac{\alpha_{Zn^{2+}} \cdot \alpha_{Cu}}{\alpha_{Cu^{2+}} \cdot \alpha_{Zn}} \quad (2-7-1)$$

The activity of pure solid is equal to 1

$$\alpha_{Zn} = \alpha_{Cu} = 1 \quad (2-7-2)$$

Substituting the above formula,

$$E = E^\ominus \frac{RT}{2F} \ln \frac{\alpha_{Zn^{2+}}}{\alpha_{Cu^{2+}}} \quad (2-7-3)$$

Where $E^\ominus$ is the electromotive force of galvanic cell when the activity $\alpha_{Zn^{2+}}$ of the zinc ion in the solution and the activity $\alpha_{Cu^{2+}}$ of the copper ion are both equal to 1.

Method 2: First determine the electrode potential according to the Nernst equation

Zn electrode potential: $\quad \varphi_- = \varphi^\ominus_{Zn^{2+},Zn} - \frac{RT}{2F} \ln \frac{\alpha_{Zn}}{\alpha_{Zn^{2+}}} \quad (2-7-4)$

Cu electrode potential: $\quad \varphi_+ = \varphi^\ominus_{Cu^{2+},Cu} - \frac{RT}{2F} \ln \frac{\alpha_{Cu}}{\alpha_{Cu^{2+}}} \quad (2-7-5)$

Where $\varphi^\ominus_{Cu^{2+},Cu}$ and $\varphi^\ominus_{Zn^{2+},Zn}$ are the standard electrode potentials of the copper and zinc electrodes when $\alpha_{Zn^{2+}} = \alpha_{Cu^{2+}} = 1$ (see Appendix).

Then, based on $\quad E = \varphi_+ - \varphi_- \quad (2-7-6)$

In the measurement of galvanic cell electromotive force, in order to make galvanic cell reaction proceed under thermodynamic reversible conditions (current $I \to 0$), it can not be measured with a voltmeter, and it is measured with a potentiometer. The principle of measuring the electromotive force of galvanic cell by the potentiometer is that the potentiometer provides an electromotive force equal to the size of the battery to be tested and opposite in direction, thereby offsetting the electromotive force of galvanic cell to be tested, thereby causing the current in the circuit to go to zero. The indication is the electromotive force of galvanic cell under test.

In this experiment, we want to measure the electromotive force of Zn-Cu, calomel - Cu, and Zn-calomel batteries. The calomel electrode used is a saturated calomel electrode, and its structure and characteristics are in the theoretical textbook.

【Apparatus and Reagents】

SDC – II digital potential difference tester, 1; Standard battery, 1; Saturated calomel electrode, 1; 250 mL Three-necked flask, 1; 50 mL beakers, 3; $CuSO_4 \cdot 5H_2O$ (AR), $ZnSO_4 \cdot 7H_2O$ (AR), NaOH(AR).

【Procedures】

1. Preparation of electrodes

(1) Copper electrode

The surface of the copper sheet was first sanded with sandpaper to remove the oxide layer on the surface, and then washed with water, and the treated Cu electrode was inserted into a clean small beaker containing 0.1 mol/L of $CuSO_4$ solution.

(2) Zinc electrode

The surface of the zinc sheet was first sanded with sandpaper to remove the oxide layer on the surface, and then washed with water, and the treated Zn electrode was inserted into a clean small beaker containing 0.1 mol/L of $ZnSO_4$ solution.

(3) Saturated calomel electrode

Observe whether there is KCl crystal in the calomel electrode. If not, add a small amount of KCl solid from the side nozzle to ensure that the saturated KCl solution can immerse the mercury at the upper end of the electrode, and there are no bubbles in the electrode.

(4) The combination of batteries

Approximately 20 mL of a concentration of 0.1 mol/L $CuSO_4$ and 0.1 mol/L $ZnSO_4$ solution were placed in a small beaker, and two beakers were connected by agar-saturated potassium chloride salt bridge, and then the copper electrode and the zinc electrode prepared above were placed. Placed in a small beaker containing $CuSO_4$ and $ZnSO_4$, respectively, into a Zn – Cu battery:

$$Zn(s) \mid ZnSO_4(0.1\ mol \cdot L^{-1}) \parallel CuSO_4(0.1\ mol \cdot L^{-1}) \mid Cu(s)$$

The same battery is composed of the same method:

$$Zn(s) \mid ZnSO_4(0.1\ mol \cdot L^{-1}) \parallel KCl(饱和) \mid Hg_2Cl_2(s) \mid Hg$$

$$Hg \mid Hg_2Cl_2(s) \mid KCl(饱和) \parallel CuSO_4(0.1\ mol \cdot L^{-1}) \mid Cu(s)$$

2. Measurement of battery electromotive force

The control panel of the potentiometer is shown in Figure 2 - 7 - 1.

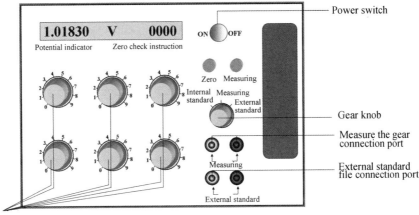

Figure 2 - 7 - 1 Control panel of SDC - II digital potential difference comprehensive tester

(1) Turn on the power switch.

(2) Rotate the gear knob to the external standard file.

(3) Connect the standard battery to the "external standard wiring terminal" of the potentiometer with a wire.

(4) Adjust the potential adjustment knob so that the output potential is 1. 018 30 V (the temperature correction formula of the standard battery electromotive force: $E_t = 1.018\ 30 - 0.000\ 040\ 6 \times (t - 20)$), observe the zero detection window display, if it is not zero, press the "return to zero" button to reset it to zero.

(5) Turn the gear knob to the measurement file.

(6) Connect the battery to be tested to the "measurement terminal" (note the positive and negative terminals of the battery).

(7) Adjust the "potential adjustment knob" according to the theoretical electromotive force of the battery to be tested. When the output potential is equal to the true electromotive force of the battery to be tested, the indication of the zero indication window is zero, and the indication of the "potential indication" window is The electromotive force of the battery.

【Records and Analysis of Data】

The theoretical values of the measured battery electromotive force were calculated according to Equations (2 - 7 - 3), (2 - 7 - 4), (2 - 7 - 5) and (2 - 7 - 6), and are listed in Table 2 - 7 - 1. When calculating, the material activity should be calculated as follows:

$$\alpha_{Zn^{2+}} = \gamma_\pm c_{Zn^{2+}}, \alpha_{Cu^{2+}} = \gamma_\pm c_{Cu^{2+}} \qquad (2 - 7 - 7)$$

Where is the average ion activity coefficient of the ions, the values of which are shown in Table 2 - 7 - 2.

Table 2-7-1 Electromotive force measurement data results

	Battery	$E_{\text{Measurements}}$ /V	$E_{\text{Theoretical value}}$ /V	Relative error/%
1	Zn(s) ∣ ZnSO$_4$(0.1 mol·L^{-1}) ∥ CuSO$_4$(0.1 mol·L^{-1}) ∣ Cu(s)			
2	Zn(s) ∣ ZnSO$_4$(0.1 mol·L^{-1}) ∥ KCl(饱和) ∣ Hg$_2$Cl$_2$ ∣ Hg			
3	Hg ∣ Hg$_2$Cl$_2$ ∣ KCl(饱和) ∥ CuSO$_4$(0.1 mol·L^{-1}) ∣ Cu(s)			

Table 2-7-2 Activity coefficients of CuSO$_4$ and ZnSO$_4$ at different concentrations

Electrolyte	Concentration	
	0.1 mol·L^{-1}	0.01 mol·L^{-1}
CuSO$_4$	0.16	0.40
ZnSO$_4$	0.15	0.387

【Notes】

The measurement method of the electromotive force belongs to the balance measurement, and it is performed under the reversible conditions as much as possible during the measurement process. Pay attention to the following points for this:

1. The measurement needs to calculate the theoretical electromotive force of the battery so that the equilibrium point can be quickly found during the measurement, so that the polarization of the electrode can be avoided.

2. In order to judge whether the measured electromotive force is a balanced potential, generally 7 to 8 data should be measured at intervals of 15 min or so. If the data is oscillated near the average value and the deviation is less than ± 0.0005 V, it can be considered that the equilibrium has been reached, and the average value can be taken as the electromotive force of the battery.

3. Connect the line and pay attention to the correct use of the positive and negative poles and the salt bridge.

【Post Lab Questions】

1. Why does the electromotive force use a potentiometer? Is it possible to use a multimeter?

2. Why is the use of "salt bridge" for measuring the electromotive force?

3. What is the main principle of measuring the electromotive force of Galvanic Cells by the elimination method?

实验 8　表面活性剂临界胶团浓度的测定

【实验目的】

1. 了解表面活性剂的性质和胶束形成的原理；
2. 掌握最大气泡压力法测定溶液表面张力的原理和方法；
3. 测定表面活性剂浓度和表面张力的关系；
4. 测定月桂酸钠的临界胶团浓度。

【预习要求】

1. 了解表面张力、表面自由能的概念及物理意义；
2. 了解表面活性剂的基本性质。

【实验原理】

由具有明显"两亲"性质的分子组成的物质称为表面活性剂。这类分子既含有亲油的足够长的烃基（一般大于 10 个碳原子），又含有亲水的极性基团。在化学结构上，都是由非极性的、亲油（疏水）的长链碳氢基和极性的、亲水（疏油）的基团共同构成的，而形成不对称的结构，如肥皂和各种合成洗涤剂等。若按亲水基团的离子性或非离子性，表面活性剂可分为阴离子表面活性剂、阳离子表面活性剂、非离子表面活性剂以及两性表面活性剂等。

表面活性剂溶入水中后，在低浓度时溶液中的表面活性剂分子呈分散状态，并且互相把亲油基团聚拢而分散在水中。当溶液浓度增大到一定程度时，许多表面活性物质的分子立刻结合成很大的聚集体，形成"胶束"。以胶束形式存在于水中的表面活性物质是比较稳定的。表面活性物质在水中形成胶束所需的最低浓度称为临界胶团浓度，以 CMC（critical micelle concentration）表示。在 CMC 点上，由于溶液的结构发生变化，导致其物理性质及化学性质（表面张力、电导、渗透压、浊度、光学性质等）发生明显变化，如图 2-8-1 所示，这个现象是测定 CMC 的实验依据，也是表面活性剂的一个重要特征。

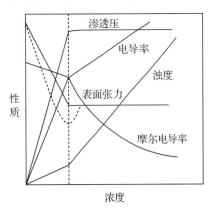

图 2-8-1　十二烷基硫酸钠水溶液物理性质与浓度关系图

这种特征行为可用生成分子聚集体或胶束来说明，如图 2-8-2 所示。当表面活性剂溶于水中后，不但定向地吸附在水溶液表面，而且达到一定浓度时还会在溶液中发生定向排列而形成胶束，表面活性剂为了使自己成为溶液中的稳定分子，有可能采取两种途径：一是把亲水基留在水中，亲油基伸向油相或空气；二是让表面活性剂的亲油基团相互靠在一起，以减小亲油基与水的接触面积。前者就是表面活性剂分子吸附在界面上，其结果是降低界面张力,形成定向排列的单分子膜,后者就形成了胶束。由于胶束的亲水基方向朝外,与水分子相互吸引,使表面活性剂能稳定地溶于水中。

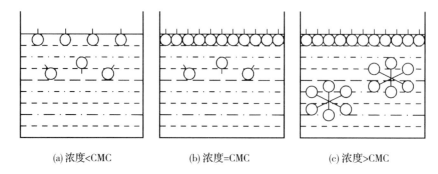

(a) 浓度<CMC　　　　(b) 浓度=CMC　　　　(c) 浓度>CMC

图 2-8-2　胶束形成过程示意图

随着表面活性剂在溶液中浓度的增长,球形胶束还可能转变成棒形胶束,甚至层状胶束,如图 2-8-3 所示。后者可用来制作液晶,它具有各向异性的性质。

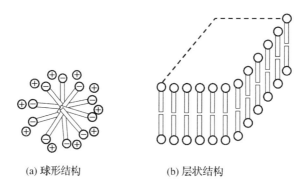

(a) 球形结构　　　　(b) 层状结构

图 2-8-3　胶束的球形结构和层状结构示意图

测定表面活性剂临界胶束浓度的方法有表面张力法、电导法及渗透压法等。本实验采用最大气泡压力法测定表面活性剂溶液的表面张力,从而得到该表面活性剂的临界胶束浓度。其实验装置如图 2-8-4 所示。

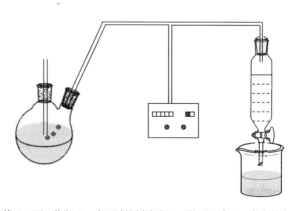

1—毛细管;2—两口烧瓶;3—表面活性剂溶液;4—微压差计;5—滴液漏斗;6—烧杯。

图 2-8-4　测定表面张力的实验装置图

将待测液体装入表面张力仪中,使玻璃毛细管下端与液面相切,若液体能润湿管壁,则液体沿毛细管上升形成凹液面。打开抽气瓶(滴液漏斗)的活塞缓缓放水抽气,此时测定管中的压力 p 逐渐减小,毛细管中的大气压力 p_0 就会将管中液面压至管口,并形成气泡。当此压力差在毛细管上面产生的作用稍大于毛细管口液体的表面张力所能产生的最大作用时,气泡就从毛细管口逸出。气泡逸出前能承受的最大压力差为 Δp_{max},可用微压差测量仪测出。根据拉普拉斯(Laplace)方程,毛细管内凹液面的曲率半径恰好等于毛细管半径 r 时(此时压力为 p_r),能承受的压力差为最大,则有

$$\Delta p_{max} = p_0 - p_r = \frac{2\sigma}{r} \tag{2-8-1}$$

$$\sigma = \frac{r}{2} \Delta p_{max} \tag{2-8-2}$$

测定毛细管的半径 r 和 Δp_{max} 即可求得液体表面张力 σ。直接测定毛细管的半径 r 容易带入较大的误差,可用同一支毛细管,在相同条件下分别测出已知表面张力为 σ_1 的参考液体的 $\Delta p_{max,1}$ 和待测液体的 Δp_{max},由式(2-8-1)得

$$r = \frac{2\sigma}{\Delta p_{max}} = \frac{2\sigma_1}{\Delta p_{max,1}}$$

$$\sigma = \frac{\sigma_1}{\Delta p_{max,1}} \Delta p_{max} = K' \Delta p_{max} \tag{2-8-3}$$

K' 称为毛细管常数,可用已知表面张力的物质来确定。由式(2-8-3)可求出待测液体的表面张力 σ。

【仪器与试剂】

微压差测量仪(DMP-2B 型),表面张力测量玻璃仪器,恒温水浴,容量瓶(50 mL),烧杯,洗瓶,月桂酸钠水溶液。

【实验步骤】

1. 配制溶液。

精确称取月桂酸钠,用重蒸馏水准确配制 0.005 mol·L^{-1}、0.010 mol·L^{-1}、0.015 mol·L^{-1}、0.020 mol·L^{-1}、0.025 mol·L^{-1}、0.030 mol·L^{-1} 及 0.035 mol·L^{-1} 的月桂酸钠溶液各 50 mL 待用。

2. 调节恒温水浴到 25 ℃(或 30 ℃)。

3. 测定毛细管常数。

将玻璃器皿洗涤干净,在烧瓶中注入蒸馏水,使液面与毛细管口相切,检查无漏气后,置于恒温水浴内恒温 10 min。然后慢慢打开抽气瓶活塞,使水滴滴下,以每秒 1~2 滴为宜,从微压差测量仪上读取压力差最大值 Δp_{max},读数三次,计算时取平均值。

4. 测定月桂酸钠溶液的表面张力。

以水为对照组按实验步骤 3 分别测量不同浓度月桂酸钠溶液的表面张力,从稀到浓依次进行。每次测量前必须用少量被测液洗涤烧瓶与毛细管,尤其是毛细管部分,确保毛细管

内外溶液的浓度一致。具体做法是:用吸耳球从毛细管上端吸、放液体三次,每次液体高度距液面约 2 cm 为宜。每次更换溶液后,在表面张力仪中应恒温 10 min 以上,为了节省时间,可先将盛待测溶液的容器放入恒温水浴内恒温,恒温后溶液放入表面张力仪后再恒温 3 min,然后进行测量。

5. 实验完毕,洗净玻璃容器。

【实验记录与数据处理】

1. 将所测的数据和结果按表 2-8-1 形式列出。

室温:_____ ℃　　　　　气压:_____ kPa

室验温度:_____ ℃　　　实验温度下水的表面张力:_____ mN/m

表 2-8-1　表面张力实验数据

溶液浓度/ (mol·L^{-1})	次数	Δp_{max}/Pa	Δp_{max}平均值/Pa	表面张力 σ/(mN·m^{-1})
水	1			
	2			
	3			
0.005	1			
	2			
	3			
0.010	1			
	2			
	3			
0.015	1			
	2			
	3			
0.020	1			
	2			
	3			
0.025	1			
	2			
	3			
0.030	1			
	2			
	3			
0.035	1			
	2			
	3			

2. 作表面张力与溶液浓度的关系图($\sigma - C$ 图),并从图中转折点处找出临界胶团浓度。

【讨论与说明】

1. 仪器调零时,要与大气相通;
2. 实验前要将表面张力仪及毛细管清洗干净;
3. 每次测量时,保持毛细管伸入液面同一深度,保持滴液漏斗内水的高度相同;
4. 换待测液时,要将原毛细管内的液体吹出,并用待测液润洗毛细管下端三次。

【思考题】

1. 实验前为什么一定要确保表面张力仪和玻璃毛细管洁净?
2. 为什么不直接测量毛细管的半径,而用标定的办法?
3. 溶解的表面活性剂分子与胶束之间的平衡同温度和浓度有关,其关系式可表示为

$$\frac{\mathrm{d}\ln C_{\mathrm{cmc}}}{\mathrm{d}T} = -\frac{\Delta H}{2RT^2}$$

试问如何测出其热效应 ΔH 值?

【相关阅读】

1. 液体表面张力的测定方法

液体表面区的分子由于受力不平衡产生的向内收缩的单位长度的力即表面张力。它分为静态表面张力和动态表面张力。通常液体的表面张力自其液体表面形成之后,随着时间的推移而有所变化。在新的液体表面形成的瞬间,经过约 1 s 以上时间的表面张力称作静态表面张力;在 1 s 以下的表面张力称作动态表面张力。表面张力是多相系统的重要界面性质,对于泡沫分离、蒸馏、萃取、乳化、吸附、润湿等过程存在重要影响。在实际生产过程中,动态表面张力更有意义,因为它反映出传质过程以及吸附、黏附、铺展等过程的有关信息,这对于化工过程的设计与研究是非常有意义的。

液体表面张力的测定方法分为静力学法和动力学法。静力学法包括毛细管上升法、du Noüy 环法、Wilhelmy 片法、旋滴法、悬滴法、滴体积法、最大气泡压力法;动力学法包括振荡射流法、毛细管波法。其中毛细管上升法和最大气泡压力法不能用来测液-液界面张力;Wilhelmy 片法、最大气泡压力法、振荡射流法、毛细管波法可以用来测定动态表面张力。由于动力学法本身较复杂,测试精度不高,而先前的数据采集与处理手段都不够先进,致使此类测定方法成功应用的实例很少。因此,迄今为止,实际生产中多采用静力学测定方法。

(1) 毛细管上升法

①测定原理

将一支毛细管插入液体中,液体将沿毛细管上升,上升到一定高度后,毛细管内外液体将达到平衡状态,液体就不再上升了。此时,液面对液体所施加的向上的拉力与液体总向下的力相等,则

$$\gamma = \frac{1}{2}(\rho_l - \rho_g)ghr\cos\theta \tag{2-8-4}$$

式中,γ 为表面张力;r 为毛细管的半径;h 为毛细管中液面上升的高度;ρ_l 为测量液体的密度;ρ_g 为气体的密度(空气和蒸气);g 为当地的重力加速度;θ 为液体与管壁的接触角。若毛细管管径很小,而且 $\theta = 0°$ 时,则式(2-8-4)可简化为

$$\gamma = \frac{1}{2}\rho ghr \qquad (2-8-5)$$

②优点

本法是用来直接测定液体表面张力的最准确的绝对方法之一,也是应用最多的方法之一。它不仅理论完整,而且实验条件可以严格控制,是一种重要的测定方法。随着技术的发展,毛细管上升技术也可以用来测定动态表面张力。此方法还曾被用于高温、高压条件下表面张力的测定,但温度一般不超过 100 ℃,压强不超过 13.8 MPa。

③缺点

a. 不易选得内径均匀的毛细管和准确测定内径值;

b. 液体与管壁的接触角不易测量;

c. 溶液的纯度会对表面张力的测量造成不同程度的影响;

d. 需要较多液体才能获得水平基准面(一般认为液面直径在 10 cm 以上才能被看作平面),所以基准液面的确定可能产生误差。

(2) Wilhelmy 片法

①测定原理

用铂片、云母片或显微镜盖玻片挂在扭力天平或链式天平上。测定当片的底边平行面刚好接触液面时的压力。由此得表面张力的公式为

$$W_{总} - W_{片} = 2\gamma l \cos\theta \qquad (2-8-6)$$

式中,$W_{总}$ 为薄片与液面拉脱时的最大拉力;$W_{片}$ 为薄片所受的重力;l 为薄片的宽度,薄片与液体接触的周长近似为 $2l$;θ 为薄片与液体的接触角。

②方法特点

此法是 1863 年由 Wilhelmy 提出来的,后来 Dognon 和 Arbribut 对其进行了改进,它具有完全平衡的特点。该方法是常用的实验方法之一,设备简单,操作方便,不需要密度数据,直观可靠,不仅可用于测定气-液表面张力,也可用于测定液-液界面张力。精确度可达到 $0.11 \text{ mN} \cdot \text{m}^{-1}$。该方法存在的缺点是:

a. 要求液体必须很好地湿润薄片,保持接触角为零;

b. 需要标准物质校正浮力;

c. 测定容器需要足够大;

d. 不适合高温高压和深颜色液体的测定;

e. 清洁程序复杂;

f. 测定时稳定慢,不适合及时测量。

③发展与变迁

a. Padayt 等提出了用湿润的棒代替薄片,所得的结果同其他方法完全一致,精度可以

达到 ± 0.1 mN·m^{-1},是一种对所有方位都平衡的方法。

b. Padday 提出了用圆锥代替薄片,并计算了圆锥常数,但并未给出较明细的装置示意图。周怀申等不仅较详细地介绍了圆锥最大拉力法的装置示意图,而且不断改进,使之具有更高的稳定性,并能用于较高温度条件下易挥发溶液的表面张力的测定,精度为 ± 0.10 mN·m^{-1}。圆锥最大拉力法是绝对法,不需要对标准物质进行校正,只要圆锥符合要求,且锥面与弯月面形成的接触角保证为零度,就可测准溶液的表面张力。

c. Buboltz 等提出了用不湿润但有光滑底部的聚四氟乙烯棒来测定。此法仍需样品完全湿润棒的底部,可用低表面能的底物与样片接触。因此适用于易破裂的表面,也有利于研究达到平衡较慢的体系。

d. Christian 等提出了在薄片或盘向上提起时测量弯月面液相的重力。此法可广泛应用于纯液体,以及阴、阳离子和中性的表面活性剂溶液表面张力的测定。

（3）悬滴法

①测定原理

悬滴法是根据在水平面上自然形成的液滴形状计算表面张力。在一定平面上,液滴形状与液体表面张力和密度有直接关系。由 Laplace 公式,任意的一点 P 曲面内外压差为

$$\gamma\left(\frac{1}{R_1}+\frac{1}{R_2}\right)=P_0+(\rho_1-\rho_g)gz \qquad (2-8-7)$$

式中,R_1、R_2 为液滴的主曲率半径;z 为以液滴顶点 O 为原点,液滴表面上 P 的垂直坐标;P_0 为顶点 O 处的静压力。

定义 $S=\dfrac{\mathrm{d}s}{\mathrm{d}e}$,其中 $\mathrm{d}e$ 为悬滴的最大直径,$\mathrm{d}s$ 为离顶点距离为 $\mathrm{d}e$ 处悬滴截面的直径;再定义 $H=\beta\left(\dfrac{\mathrm{d}e}{b}\right)^2$,其中,$b$ 为液滴顶点 O 处的曲率半径;β 为常数。则得

$$\gamma=\frac{(\rho_1-\rho_g)g\mathrm{d}e^2}{H} \qquad (2-8-8)$$

此式最早是由 Andreas、Hauser 和 Tucker 提出的,若 $1/H$ 为已知,即可求出表(界)面张力。应用 Bashforth - Adams 法,即可算出作为 S 的函数的 $1/H$ 值。因为可采用定期摄影或测量 $\mathrm{d}s/\mathrm{d}e$ 数值随时间的变化,悬滴法可方便地用于测定表(界)面张力。

②优点

该方法除了对样品的湿润性无严格要求、不受接触角影响外,还有测定范围广(不仅可测定液体的静态,还可测定液体的动态表面张力)的特点。这是一种液体用量少而且应用广泛的方法,也比较适用于高温高压条件下液体表面张力和低表面张力的测定,可以用来测定 200 ℃和 81.7 MPa 条件下的液体表面张力。

③缺点

a. 设备复杂,操作麻烦;

b. 数据处理复杂;

c. 待测物质的性质需要事先准确知道。

(4) 滴体积法

①测定原理

当一滴液体从毛细管滴头滴下时,液滴的重力与液滴的表面张力以及滴头的大小有关。Tate 首先提出了表示液滴重力(mg)的简单关系为

$$mg = 2\pi r\gamma \tag{2-8-9}$$

实验结果表明,实际体积比按式(2-8-6)计算的体积小得多。因此 Harkins 就引入了校正因子 $f(r/V^{1/3})$,则更精确的表面张力可以表示为

$$\gamma = \frac{mg}{2\pi r f(r/V^{1/3})} \tag{2-8-10}$$

式中,m 为液滴的质量;V 为液滴体积;f 为校正因子,可查表得到。

只要测出数滴液体的体积,利用式(2-8-10)就可计算出该液体的表面张力。

②方法特点

此法是一种相对精确而又可能是最方便的方法之一,它的样品制备简单,温度时间间隔长,只用简单的温度控制即可,可用来测定气-液和液-液界面,且样品的用量少。存在的缺点如下:

a. 至今只能算是一种经验方法;

b. 不能用来测定达到平衡较慢的表面张力,同时该法也不能达到完全的平衡;

c. 存在准确测定液体体积和很好地控制液滴滴落速度等问题。

③发展与变迁

Kevin Hool 等提出:测量毛细管尖端的几何构型对于测量非常重要。因此可用精确的定气体流量源来产生气泡,然后用精确的计数器或者计时器来测量气泡分离的时间。这样就使得该装置有更大的灵活性来测定液-液界面张力。Hitoshi Matsuki 等采用了电动千分尺和光电传感器可自动获得具有足够准确性和重复性的表面张力。

(5) 最大气泡压力法

①测定原理

若在密度为 ρ 的液体中插入一个半径为 r 的毛细管,深度为 t,经毛细管吹入一极小的气泡,其半径恰好与毛细管半径相等,此刻气泡内压力最大。根据拉普拉斯公式气泡最大压力为

$$p_m = \rho g t + \frac{2\gamma}{r} \tag{2-8-11}$$

即

$$\gamma = 1/2 r(p_m - \rho g t) \tag{2-8-12}$$

②优点

这是由 Simon 于 1851 年提出的方法,后来由 Canter 和 Jaeger 分别从理论和实用角度加以发展。此种方法设备简单,操作方便,不需要完全湿润,它既是相对的方法,也是绝对的方法,可以测量静态和动态的表面张力,测量的有效时间范围大,温度范围宽。

③缺点

a. 气泡不断生成可能会扰动液面平衡,改变液体表面温度,因此不易控制气泡形成速度;

b. 要求在气泡逸出瞬间读取气泡的最大压力,因此此值很难测准;

c. 毛细管的半径不易准确测定;

d. 此法中,最大压差为大气压与系统压力的差值,因此,当室内气流流动时,会造成大气压的变化,使实验测得的数据产生一定误差;

e. 为了消除溶液静压对测定结果的影响,测定时要求测量的毛细管插入液体中的深度为0,但要调整毛细管尖端与被测液面相切有一定的难度。

2. 临界胶束浓度的测定方法

(1) 电导法

这是测定临界胶束浓度(简称CMC)的经典方法,所用仪器为各种型号的电导率仪,并可与计算机配套使测定自动化。原则上讲,此法最适用于离子型表面活性剂CMC的测定,且对有较高表面活性的表面活性剂准确性高。但在一定条件下此法也可用于非离子型表面活性剂CMC的测定。

对于离子型表面活性剂,当溶液浓度很稀时,电导的变化规律与一般强电解质相似,表面活性剂完全解离为离子,随着浓度上升电导率K近乎线性上升。但当溶液浓度达到临界胶束浓度时,随着胶束的形成,胶束定向移动速率减缓,电导率K仍随着浓度增大而上升,但变化幅度变小,摩尔电导率λ_m也急剧下降,可利用$K-C$曲线或λ_m-V_c曲线的转折点求CMC值。

(2) 表面张力法

表面活性剂溶液的表面张力随浓度的变化在CMC处同样出现转折。只要测定了不同浓度表面活性剂溶液的表面张力,由$\gamma-C$关系曲线可方便地求得CMC。测液体表面张力的方法前面已叙述。

(3) 染料吸附法

此法是利用某些染料在水中和胶束中的颜色有明显差别的性质,采用滴定的方法测定CMC_o。根据同性电荷相斥、异性电荷相吸的原理,所用染料(一般为有机离子)必须要与表面活性离子的电荷相反。故此法的关键是选择合适的染料。如常用于阴离子(负离子)表面活性剂的染料有频那氰醇氯化物、碱性蕊香红G等;用于阳离子(正离子)表面活性剂的染料有曙红、荧光黄等;用于非离子型表面活性剂的染料有四碘荧光素、苯并红紫4B等。其具体操作的方法是先在一确定浓度(>CMC)的表面活性剂溶液中加入少量的染料,此时染料被溶液中的胶束吸附而使溶液呈现某种颜色,再用滴定的办法用水冲稀此溶液,直至溶液颜色发生显著变化,由被滴定溶液的总体积可方便求得CMC。如氯化频那氰醇在CMC以上时为蓝色,CMC以下为红色。由于此法是以颜色的变化来确定CMC值的,所以此方法对溶液本身颜色较深的体系并不适用;若体系含盐或醇较多时,会导致颜色变化不明显,则此法也不适用。此外,该法用于非离子型表面活性剂测定效果也不甚理想,遇到这些情况可考虑采用其他方法如表面张力法、紫外吸收光谱法等来测定。

(4) 紫外吸收光谱法

前述染料吸附法用于非离子型表面活性剂CMC测定,效果很不理想。此时改用微量的碘代替染料,在紫外区适宜的波长下观察光谱的变化可提高测定灵敏度。此法的关键在于不同的体系需选用不同的测定波长。

(5) 溶解度法

表面活性剂的加入使得某些不溶或微溶于水的有机化合物的溶解度显著增加。研究表明,被增溶的物质(如碳氢化合物)溶解于胶团内部增水基团集中的地方,故测定溶液的溶解度 S,由 $S-C$ 关系曲线上出现的转折点可方便地求得 CMC。例如,2-硝基二苯胺在月桂酸钾水溶液中的溶解度,当月桂酸钾浓度达 0.5% 时会急剧增加。此法在实际测定中可采用不同的操作手段。其一为直接测定溶解度;其二为间接法,即用颜色的变化来求溶解度,进而确定 CMC。方法是在体系中加入不溶于水的固体染料,然后由小到大改变表面活性剂的浓度。当达 CMC 时染料的溶解度急剧增加,则整个溶液呈现染料的颜色。这种方法非常适用于以水为溶媒的体系中表面活性剂的 CMC 测定。但是要求体系本身颜色较浅。此外染料的加入会对 CMC 较小的表面活性剂 CMC 的测定有影响,遇到这些情况可选用烃类代替染料。因为在稀表面活性剂溶液(<CMC)中,烃类一般不溶或不随浓度改变,但当表面活性剂浓度超过 CMC 后溶解度急剧增加。一般可由溶液的浊度变化来确定溶解度。

(6) 光散射法

光散射现象是溶胶体系特有的光学性质,即胶体化学中著名的丁达尔效应。由于 CMC 是表面活性剂溶液的物理化学性质发生跃变的"分水岭",所以当表面活性剂在溶液中达到或超过一定浓度时,会从单体(单个离子或分子)缔合成胶态聚集物即形成胶团,其大小符合胶粒大小的范围故有光的散射现象。随着表面活性剂浓度的增大,胶团的聚集数不断增多,光散射强度随之增强,当达 CMC 时,光散射强度急剧增加,CMC 即可由光散射强度-浓度图中的突变点求出。此法的特点在于除获得值外,还可测定胶团的聚集数、胶团的形状和大小及胶团的电荷量等有用的数据,这是上述其他方法不易办到的。此外,该法要求待测溶液非常纯净,任何杂质质点都将影响测定结果。

Experiment 8　Determination for Critical Micelle Concentration of Surfactant

【Purpose of the Experiment】

1. To determine critical micelle concentration (CMC) of an anionic surfactant, sodium dodecyl sulphate.

2. To learn the principles and technique of maximum bubble pressure method for measuring critical micelle concentration (CMC) of surfactant.

【Principle】

Surfactant is an amphiphilic substance that consists of both polar and non-polar parts. When a small amount of surfactant is dissolved in water, it mainly exists as monomers dispersed in water. At the surface of the aqueous solution, polar groups point to the interior of the solution, and non-polar groups to the atmosphere. The arrangement of surfactant molecules at the surface varies with the concentration. When the concentration of the surfactant is low, the surfactant molecules tend to lie down on the surface. With the increase of the surfactant concentration, the non-polar group of the molecule stands up gradually until a saturated adsorption layer is formed. After that, as the surfactant concentration is increased, the monomers form aggregates called "micelles". The minimum concentration of surfactant at which micelle formation begins is referred to as the critical micelle concentration, abbreviated to CMC which is a very important property for surfactants. The formation of micelles is shown in Figure 2 – 8 – 1.

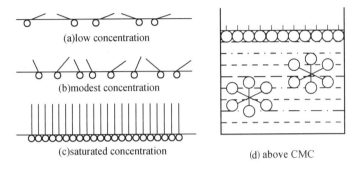

Figure 2 – 8 – 1 The formation of micelles

Near the CMC, the physicochemical properties (such as surface tension, conductance, osmotic pressure, optical property) change abruptly with concentration owing to the CMC formation in solution, as shown in Figure 2 – 8 – 2. Only above CMC a surfactant can take effect in wetting, emulsion, washing, foaming, etc. The abrupt change of properties of surfactants is the base of CMC measurement, these methods include the measurement of surface tension, conductance, turbidity, light scattering, etc. This experiment uses surface tension method (i. e. maximum bubble pressure method) to determine the CMC of sodium dodecyl sulphate. The relation of surface tension σ with concentration c is shown in Figure 2 – 8 – 3. The CMC can be obtained from the curve of $\sigma - c$.

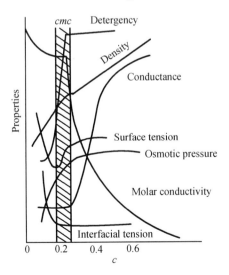

Figure 2 – 8 – 2 Physicochemical properties of surfactant solutions

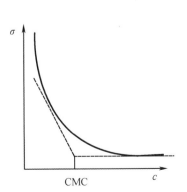

Figure 2 – 8 – 3 The relation of surface tension σ with concentration c

Maximum bubble-pressure method is one of most popularly used techniques for measuring the surface tension of liquids. The experimental apparatus is shown in Figure 2-8-4. When the bottom of a capillary tube is just in contact with the surface of the test liquid, the liquid will rise up in the capillary. Slowly open the stopcock of the dropping funnel to discharge water from the bottle. The pressure in the test tube P_r is reduced accordingly, so the atmospheric pressure P_0 in the capillary is greater than P_r to cause the formation of bubbles.

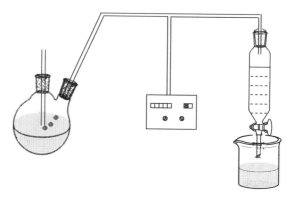

Figure 2-8-4 The schematic diagram of the apparatus for measurement of surface tension by maximum bubble pressure method

According to the Laplace equation, when the curvature radius of the bubble equals the radius of the capillary, a maximum pressure difference between P_0 and P_r described as follow:

$$\Delta P = \frac{2\sigma}{r}$$

Where ΔP——excess pressure, Pa;

σ——surface tension of solution, N·m^{-1};

r——curvature of bubble, m.

When the dropping continues, P_0 will expel the bubble out of the capillary, which leads to a further increase of the curvature radius of the bubble and a reduction in the endurable pressure difference of the bubble. As the pressure difference in the test tube becomes even larger, the bubble will break. The maximum pressure difference can be read from the digital manometer.

According to the above equation, the following relationship can be derived

$$\sigma = \frac{r}{2}\Delta p = K\Delta p$$

Where K is the capillary constant which can be determined with substance of known surface tension. If we use a known surface tension sample as standard, the apparatus constant K can be obtained after ΔP was measured from experiment. The surface tension σ of unknown sample can be determined using the same apparatus and method. In this experiment, distilled water is used as standard, which surface tension at 20 ℃ is 72.75 mN·m^{-1}.

【Apparatus and Reagents】

The surface tension measurement apparatus; six volumetric flasks (50 mL); sodium dodecyl sulphate; distilled water.

【Procedures】

1. Preparation of various concentrations of sodium dodecyl sulphate solutions

Preparing various concentrations (0.002 mol·L^{-1}, 0.004 mol·L^{-1}, 0.006 mol·L^{-1}, 0.008 mol·L^{-1}, 0.010 mol·L^{-1}, 0.012 mol·L^{-1}) of sodium dodecyl sulphate solutions according to general method in the laboratory.

2. Preparation and check of the apparatus

The apparatus is set up according to Figure 2-8-4, and some tap water is added to the dropping funnel. The manometer is switched on, which's unit is set to mmH$_2$O. Press the "Zero" button to make the manometer show "0".

3. Determination of the capillary constant

A suitable amount of distilled water is added to the sample tube to sure that the liquid level is just in contact with the end of the capillary which vertical. Opening the piston of dropping funnel, then water will drip slowly. Keeping the dripping speed at 1-2 drops per second. The instant maximum pressure difference should be read and recorded, and the final data should be the average value of 3 separate readings.

4. Measuring the surface tension of sodium dodecyl sulphate solutions

To measure the instant maximum pressure difference of various concentrations of sodium dodecyl sulphate solutions, respectively, according to procedure 3. It is worth noting that flasks and capillaries should be moistened with the liquid to be tested for each change in concentration.

【Notes】

1. Make sure that the system has a good sealing property and is clean.
2. The capillary must be cleaned before use.
3. The capillary plane is tangential to the level of the liquid.
4. The sequence of measuring surface tension of sodium dodecyl sulphate solutions must be from dilute concentrations to dense ones.

【Data Analysis】

1. List recording data and calculating results in the table 2-8-1.
2. Plot $\sigma - c$ curve using origin soft and obtain CMC of micelles at turning point.

Table 2-8-1 Experimental data and calculation results

Concentration of Sample (mol·L^{-1})	Δp_{max} (mmH$_2$O)			Δp_{max} (mmH$_2$O) Mean	Surface tension σ (mN·m^{-1})
	1	2	3		
Control (water)					
0.002					
0.004					
0.006					
0.008					
0.010					
0.012					

【Post Lab Questions】

1. Why must the capillary be tangential to the liquid level?
2. How would fast dropping influence the result?
3. Why should we measure the maximum bubble pressure?

Experiment 9 Determination of the Rate Law for the Iodination of Acetone

【Purpose of the Experiment】

1. To learn about the Kinetics of Chemical Reactions.
2. To learn about the Reaction Rate.
3. To learn about the Factors Affecting the Chemical Reaction Rate.
4. To learn about Rate Laws.

【Background Information】

In this laboratory exercise we will determine the Rate Law for the acid (H^+) catalyzed reaction between Acetone (CH_3COCH_3) and Iodine (I_2).

$$CH_3COCH_3(aq) + I_2(ap) \xrightarrow{H^+} CH_3COCH_2I(aq) + HI(aq) \quad (2-9-1)$$

clear sol'n   yellow sol'n   clear sol'n   clear sol'n

The catalyst H^+ is not listed as a reactant or a product, although it participates in the chemical reaction, any H^+ that is consumed is eventually regenerated.

Study of this reaction began at the turn of 20th century when it was noticed the Yellow color of an aqueous solution of Iodine (I_2) slowly fades when the solution also contains Acetone and is allowed to stand in the sunlight. Since that time, reactions involving a reactant containing a central Carbonyl Group (Figure 2-9-1):

Carbonyl Group

Figure 2-9-1 Carbonyl group structure

have become some of the most studied organic chemical reactions. The kinetics and mechanisms of these reactions are very well understood. And, frequently, reactions involving this grouping are indeed acid catalyzed.

The Rate of this reaction is easily measured due to the fact that Iodine is the only colored species (yellow in an aqueous environment) present. The faster the yellow color of the Iodine fades, the faster the reaction is occurring.

The Rate of Reaction for any general reaction:

$$aA + bB \longrightarrow cC + dD \qquad (2-9-2)$$

can be defined in terms of any observable species involved in the reaction:

$$\text{Rate of Reaction} = \frac{-1}{a}\frac{d[A]}{dt} \qquad (2-9-3)$$

$$= \frac{-1}{b}\frac{d[B]}{dt}$$

$$= \frac{+1}{c}\frac{d[C]}{dt}$$

$$= \frac{+1}{d}\frac{d[D]}{dt}$$

In practice it is difficult to determine instantaneous concentration changes, so instead we approximate the above definitions by:

$$\text{Rate of Reaction} = \frac{-1}{a}\frac{\Delta[A]}{\Delta t} \qquad (2-9-4)$$

$$= \text{etc.}$$

For the Iodination of Acetone, the observable species is Iodine, so we define the Reaction Rate as:

$$\text{Rate of Reaction} = -\Delta[I_2]/\Delta t \tag{2-9-5}$$

In general, the rate of reaction can depend on any number of factors, just as the rate of travel in an automobile depends on the position of the accelerator or brake, hilliness of the ground, wind speed, size of the engine, etc. In the case of a chemical reaction, the Rate can depend on:

  Reactant Concentrations
  Product Concentrations
  Catalyst Concentration
  Temperature
  Amount of Light
  Size of Heterogeneous Particles
  etc.

These factors will not affect the Rate of all chemical reactions, but at least some will affect any given reaction. And, occasionally, odd factors, such as such the amount of dust present in the reaction vessel, will also affect the Rate.

Typically, Reactant Concentration and Temperature are the major influences on the rate of reaction. At a given temperature, the Rate of Reaction frequently depends on reactant concentration according to:

$$\text{Rate} = k[A]^n[B]^m \tag{2-9-6}$$

where A and B are the reaction reactants. This form of the reaction rate dependence is known as the Rate Law for the reaction and k is referred to as the Rate Constant. The reaction is said to proceed according to Orders n and m. It should be emphasized the parameters k, n, and m must be measured experimentally.

For our reaction, the Rate Law will have the following form:

$$\text{Rate} = k[Ac]^n[I_2]^m[H^+]^p \tag{2-9-7}$$

Note the presence of the catalyst H^+ in the Rate Law. The reaction rate certainly depends on the presence of the catalyst, so it is not unreasonable to assume the reaction rate will depend on its concentration. It is the goal of this laboratory exercise to measure the Reaction Orders n, m and p and thus determine the Rate Law for the Iodination of Acetone.

So, how are the reaction orders for a chemical reaction measured? There are a number of methods available. One of the most common is the Method of Initial Rates. In this procedure, we selectively change the Initial Concentration of each Species and see how this change affects the rate of reaction. Suppose we double the Initial Concentration of Species A and find it causes the reaction rate to quadruple. We then know $n = 2$; or the reaction is Second Order in A. We do the same with each subsequent Species suspected of influencing the rate of reaction to determine each

Reaction Order. (The Initial Rates are used in this methodology because it provides a convenient reference point to compare one trial to another.)

For an example, consider the following reaction:

$$NO_2(g) + CO(g) \longrightarrow NO(g) + CO_2(g)$$

We will assume the Rate of this Reaction is governed by a Rate Law of the following form:

$$\text{Rate of Reaction} = k [NO_2]^n [CO]^m$$

In order to determine the Reaction Orders n and m, the following Initial Rate data has been obtained (Table 2-9-1):

Table 2-9-1 Initial rate data

Trial	Init. Rate/(mol·L^{-1}·s^{-1})	Init. [NO$_2$]/(mol·L^{-1})	Init. [CO]/(mol·L^{-1})
1	0.005	0.10	0.10
2	0.080	0.40	0.10
3	0.005	0.10	0.20

In comparing Trials 1 and 2, we see a quadrupling of the initial NO$_2$ concentration cause the Reaction Rate to increase by a factor of 16; $0.005 \times 16 = 0.080$. This means the Reaction Order $n = 2$. The reaction is Second Order with respect to NO$_2$. Now we compare Trials 1 and 3. Here a doubling of the initial CO concentration causes no change in the Reaction Rate. Hence, the Reaction Order $m = 0$ and the reaction is Zeroth order with respect to CO. Thus, the Rate Law for this reaction is:

$$\text{Rate} = k [NO_2]^2$$

Thus, for our reaction we need to selectively vary the concentrations of Acetone, Iodine and Acid and determine how the Rate changes. The key to determining the reaction orders is the ability to measure the reaction rate for each variation in initial reactant concentration. And, as was noted above, this is simply a matter of determining how fast the yellow color of the Iodine fades.

Our life is made even simpler by arranging the experimental conditions so that the Iodine concentration is always significantly less than that of the Acetone and Acid concentrations. This is done by simply starting with a great excess of both Acetone and H$^+$:

$$[I_2]_o \ll [Ac]_o \text{ or } [H^+]_o \tag{2-9-8}$$

This means the Acetone and H$^+$ concentrations do not change appreciably during the course of the reaction, hence only their initial concentrations will affect the reaction rate. And, it is frequently found reactions of this type do not depend on the concentration of the Halogen, so:

$$\text{Rate} \propto [I_2]_0 \qquad (2-9-9)$$

Thus, assuming this finding holds for this reaction and our experiment is arranged with excesses of Acetone and Acid, the reaction rate will remain constant throughout the course of the reaction, giving us a linear decrease in Iodine concentration, as shown in Figure 2-9-2.

This means:

$$\text{Rate of Reaction} = -\Delta[I_2]/\Delta t = -(0 - [I_2]_0)/(t-0) = [I_2]_0/t$$
$$(2-9-10)$$

where t represents the time required for the yellow color to faded completely; $[I_2] = 0M$. Of course, the last assumption embodied in Equation 2-9-9 must be verified experimentally.

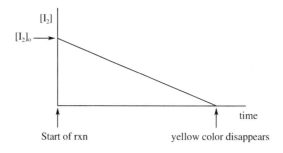

Figure 2-9-2 $[I_2]$ vs time curve

Finally, why the Rate Law has the particular form it does can only be explained by examining the reaction's Mechanism. In our case, the role of the Acid catalyst is to attack the electronegative Oxygen atom of the Carbonyl grouping:

placing an unstable Formal positive charge on this atom. The Oxygen atom rectifies this by absorbing a pair of electrons from the double bond, leaving the Carbon atom without a complete octet:

This causes a cascade of subsequent bond breaking leading to the final product. Taken together, these steps can be represented by the following Mechanism:

Step 1
Equilibrium

Step 2

Step 3
Very Fast

The validity of this Mechanism can only be confirmed by measuring the Rate Law for the reaction and determining if it is consistent with that predicted by this Mechanism.

Thus, we will determine the Rate Law for the Iodination of Acetone by measuring the Reaction Orders n, m and p using the Method of Initial Rates, obtaining the needed Reaction Rates by observing the disappearance of the characteristic Iodine color. We will then compare the experimentally determined Rate Law to the Rate Law provided by the above proposed Mechanism.

【Procedure】

Stock solutions of each of the reagents are provided at the following concentrations:

[Acetone] = 4 mol/L

[Iodine] = 0.005 mol/L

[HCl] = 1 mol/L

4 Trials of this experiment will be performed; varying the concentration, by varying the volume added, of one of the reagents, but not the others, in each case. These will be prepared as shown in Table 2-9-2:

Table 2-9-2 Each of the reagents as the following concentrations

Trial	Acetone	Iodine	HCl	Water
1	5	5	5	10
2	10	5	5	5
3	5	5	10	5
4	5	10	5	5

1. Use two large test tubes. These will be used as references for observing the end point of the reaction. Fill them with deionized Water and make sure they look clear when viewed against a

white sheet of paper.

2. Label three clean 100 mL beakers:

Acetone

HCl

Iodine

Obtain 60 mL of each stock solution and place them in the appropriate beaker.

3. Obtain a large test tube. With a pipette or graduated cylinder, measure exactly 5.00mL of the Acetone solution, 5.00 mL of the HCl solution and 10.0 mL of Water into the test tube using a pipet. Shake the test tube gently to mix. Add 5.00 mL of the Iodine solution using a pipet, start timing the reaction immediately and invert the test tube a couple of times to thoroughly mix the reactants. Note the time it takes for the solution to become clear using the test tubes from step 1 as a reference. Repeat the procedure. The two results should agree within 10 seconds. If they do not, repeat the measurement a third time.

4. Repeat this procedure for each of the remaining trials.

5. Measure the temperature of the room. All kinetic data should include a reporting of the temperature at which the reaction is proceeding as the rate of a chemical reaction usually depends on the temperature.

【Data Analysis】

1. Calculate the Initial Concentration, in units of Molarity, for each reagent used in each of the four trials in this experiment. Make an appropriate Table of these concentrations.

2. Calculate the Reaction Rate for each trial using an average of the Time required for the reaction to go to completion for each trial. Add these to you above Table.

3. Calculate the Reaction Order for each Species; m, n, p. (See the note for Pre-Lab Question 2 and follow the same procedure here.) Report two values for each Reaction Order; the exact experimentally determined value and the nearest Integer value.

4. Report the Reaction Rate Law.

【Post Lab Questions】

1. Draw Lewis Structures for Acetone (CH_3COCH_3) and Iodo Acetone (CH_3COCH_2I).

2. In the Iodination of Acetone, Iodine is Reduced. Show this by explicitly assigning Oxidation States to the Iodine in I_2 and HI.

3. Why is it necessary to keep the Total Volume of reagents in each of the four trials in this experiment the same.

4. The Rate Law for the decomposition of N_2O_5:

$$2N_2O_5(g) \longrightarrow 2N_2O_4(g) + O_2(g)$$

is found to be:

Rate = $k\,[N_2O_5]$

where $k = 0.35/\text{min}$. What is the Reaction Rate when $[N_2O_5] = 0.05$ mol/L? (Extra Credit)

5. Use the predicted Mechanism for this reaction to determine the Rate Law. Does it agree with your experimentally determined Rate Law? Explain any discrepancies.

a) First write the Rate Law as predicted by Step 3 of the Mechanism.

b) Apply the "Equilibrium" approximation to Step 1 of the Mechanism and solve for $[\text{AcH}^+]$.

c) Apply the "Steady State" approximation to Ec and solve for $[\text{Ec}]$.

d) Insert your expression for $[\text{AcH}^+]$ into the resulting equation.

e) Use your expression for $[\text{Ec}]$ in the Rate Law predicted in (a).

f) Now apply the caveat of Step 3 of the Mechanism; namely,

$$k_3[I_2] \gg k_{-2}[H^+]$$

Experiment 10　　Determination of Molar Mass by Freezing Point Depression

【Purpose of the Experiment】

The purpose of this experiment is to determine the molecular mass of organic compounds which are dissolved in a solvent by noting the depression in the freezing point of the solution as compared to the freezing point of pure solvent since freezing point depends only on the number of particles that are dissolved.

【Background Information】

Adding a solute to a pure, non-volatile liquid solvent Lowers the vapor pressure of that solvent, raises its Boiling point, and lowers (depresses) its freezing point. The extent to which these properties are affected depends on the relative number of solute particles in solution. Properties of solutions such as osmotic pressure, vapor pressure lowering, boiling point elevation, and freezing point depression, which depend only on the relative number of solute particles present, are known as colligative properties.

In this experiment, we will study the effect of solutes on the freezing point of a solvent. Freezing point depression, Δt_f, is the difference between the freezing point of the pure solvent, t_f^0, and the freezing point of the solution of the solvent and a solute, t_f,

$$\Delta t_f = |t_f^0 - t_f| \qquad (2-10-1)$$

For solutes that do not dissociate in solution (non-electrolytes), the extent of the freezing point depression depends solely on solute concentration; it is independent of the solute's chemical nature. We express solute concentration in terms of molality (m), the number of moles of solute per kilogram of solvent. We calculate molality from the solute and Solvent masses. The mathematical relationship Between freezing point depression and solute concentration for dilute, non-electrolyte solutions is represented by Equation(2 - 10 - 2). K_f is the molal freezing point depression constant of the solvent, and m is the molality of the solution.

$$\Delta t_f = K_f m \qquad (2-10-2)$$

As noted, Equation(2 - 10 - 2) applies only to dilute solutions of non-electrolytes. That is because binary strong electrolytes dissociate in solution to form two moles of solute particles per mole of electrolyte. For example, one mole of NaCl dissociates to form one mole of Na^+ (aq) and one mole of Cl^- (aq). This means that a binary strong electrolyte will be twice as effective as a non-electrolyte in depressing the freezing point of a solvent. In such cases, the value of m in Equation(2 - 10 - 2) should be doubled:

$$\Delta t_f = K_f(2m) \qquad (2-10-3)$$

The molal freezing point depression constant, K_f, differs for each solvent, but it is not dependent on the identity of the solute. Table 2 - 10 - 1 lists freezing points and molal freezing point depression constants for five solvents.

Table 2 - 10 - 1 Freezing points and molal freezing point depression constants for five solvents

Solvent	formula	freezing point/℃	K_f in ℃/mol
Water	H_2O	0.00	1.853
Benzene	C_6H_6	5.533	5.12
1,4 - dioxane	$C_4H_8O_2$	11.8	4.63
p-xylene	$C_{10}H_{10}$	13.263	4.3
carbon tetrachloride	CCl_4	-22.95	29.8

We frequently use freezing point depression data to determine the molar mass (MM, g/mol) of an unknown in solution. We can determine the molality of a solution from the freezing point depression for the solution and the K_f for the solvent. Once we know the molality of the solution, we can calculate the molar mass of the solute from the solution molality and the solvent and solute masses, using Equation(2 - 10 - 4).

$$\text{molality} = \frac{\text{moles of solute}}{\text{kilograms of solvent}} = \frac{\frac{\text{grams of solute}}{MM \text{ of solute}}}{\text{kilograms of solvent}}$$

$$MM \text{ of solute} = \frac{\text{grams of solute}}{(\text{molality})(\text{kilograms of solute})} \qquad (2-10-4)$$

For example, a student obtained the following data for an aqueous solution of an unknown solute:

mass of water, g	20.67
mass of unknown solute, g	1.04
measured freezing point of water, ℃	0.07
measured freezing point of solution, ℃	-1.50

From these data, we can calculate the molar mass of the unknown solute. We begin by using Equation(2-10-1) to determine the freezing point depression, Δt_f:

$$\Delta t_f = | t_f^0 - t_f |$$

$$\Delta t_f = | (0.07) - (-1.50) | = 1.57 \ ℃$$

We can determine the molality of the solution using Equation(2-10-2), in which K_f is the molal freezing point depression constant for water (see Table 2-10-1):

$$\Delta t_f = K_f m$$

$$1.57 \ ℃ = (1.853 \ ℃/molal) m$$

$$m = 0.847 \ mol/kg$$

Finally, we can determine the molar mass of the unknown, using Equation(2-10-5).

$$\text{MM of solute} = \frac{\text{grams of solute}}{(\text{molality})(\text{kilograms of solute})} \qquad (2-10-5)$$

$$= \frac{1.04 \ g}{(0.847 \ mol/kg)(0.020 \ 67 \ kg)}$$

$$= 59.4 \ g/mol$$

Figure 2-10-1 shows a cooling curve for a typical pure liquid solvent, such as water. An abrupt change in the slope of the curve occurs at the point where the solvent begins to freeze. As shown, many liquid solvents supercool before the solid phase begins to separate from the liquid phase. This phenomenon does not occur with every solvent. However, when it does, the solvent temperature falls below the actual solvent freezing point before crystals begin to form. Once the solvent begins to crystallize, the temperature rises abruptly, due to the energy released during crystal formation. The solution temperature then levels off at the true freezing point. The horizontal portion of the curve indicates the measured freezing point of the solvent. At this point, solid and liquid phases coexist. As we continue to remove heat from the system, more and more of the liquid freezes. Once all of the liquid has frozen, the temperature of the solid begins to drop.

When you perform this experiment, you will construct a cooling curve for pure water, based on a plot of temperature-time data for pure water. However, some of the water may not freeze, in which case your curve will not show the final temperature drop. You will also plot temperature-time data for your unknown solution, in order to construct a cooling curve for your solution. A cooling curve for a typical solution is shown in Figure 2-10-2. As the curve indicates, immediately after a solution super cools, the solution temperature starts gradually falling again.

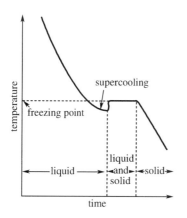

Figure 2-10-1 Cooling curve for a typical pure liquid solvent

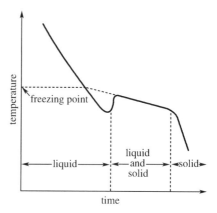

Figure 2-10-2 Cooling curve for a typical solution

That is because the solid that forms consists of pure solvent. This causes a gradual increase in solute concentration in the remaining solution, resulting in freezing point depression. We can see the difference between the behavior of a pure solvent and a solution during supercooling by comparing Figures 2-10-1 and 2-10-2.

After determining the freezing point depression for your solution, you will use these data and Equation(2-10-1) ~ Equation(2-10-4) to determine the molar mass of the unknown solvent.

【Procedure】

Preview

● Assemble freezing point apparatus.

● Prepare an ice-salt water bath.

● Record temperature-time data for freezing water.

● Determine freezing point of an unknown solution.

I. Determining the Freezing Point of Water.

1. Obtain the apparatus shown in Figure 2-10-3.

2. Clean the smaller test tube. Pour approximately 20 mL of distilled or deionized water into the tube.

3. Insert the cork holding the wire stirrer and thermometer into the smaller test tube. Make certain that the thermometer bulb is completely immersed in water but does not touch the sides or bottom of the test tube.

4. Prepare an ice-salt water bath by putting crushed ice in a layer 3~4 cm deep in a 1 L beaker. Sprinkle a thin layer of rock salt on top of the ice. Continue to alternate layers of ice and rock salt until the beaker is full. Then stir the ice-salt mixture.

5. Immerse the smaller test tube in the ice-salt water bath. Allow the water in it to cool to between 3 ℃ and 4 ℃.

6. Remove this test tube from the bath, and dry the outside. Insert the tube into the drilled-out stopper. Then insert the stopper and test tube into the larger testtube, forming an air jacket.

7. Immerse the jacketed test tube into the ice-salt water bath, and start recording temperature readings every 30 s. Estimate each temperature measurement to within 0.01 ℃. Use the wire stirrer to stir the water in the tube continuously, at a uniform rate. occasionally stir the ice-salt water bath. Record temperature-time data until a constant freezing point temperature has been reached.

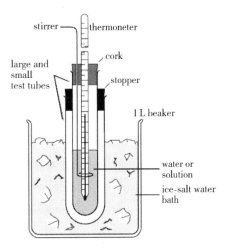

Figure 2 – 10 – 3 Freezing point determination apparatus

If you measure the freezing point of water using your assigned thermometer to be within 0.20 ℃ of 0.00 ℃, one determination is sufficient. Otherwise, make a second determination using a fresh sample of distilled water.

Ⅱ. Determining the Freezing Point of an Unknown Solution.

8. Obtain an unknown compound from your laboratory instructor. Record the identification code of your unknown. Your laboratory instructor will tell you the approximate unknown sample size to use for each determination.

9. Dry the smaller test tube thoroughly with a disposable towel, and weight to the nearest 0.01 g. Record this mass.

10. Add to this test tube the approximate mass of unknown indicated by your laboratory instructor. Reweigh the test tube and sample. Record the total mass.

11. Add about 20 mL of distilled water to the test tube, and reweight. Record the new mass.

12. Insert the stopper with the thermometer and wire stirrer into the test tube. Carefully stir the mixture in the test tube until the unknown has completely dissolved.

13. Replenish the ice-salt water bath by pouring off most of the salt water that has formed. Then add more ice and rock salt, in layers, and mix.

14. Immerse the smaller test tube in the ice-salt water bath. Allow the solution to cool to between 3 ℃ and 4 ℃. Remove the test tube from the bath, dry off the outside, and insert the test tube into the hollowed-outs topper.

15. Insert the stopper and test tube into the larger test tube. Return the jacketed test tube to the ice-salt water bath. Begin recording temperature reading severy 30 s. Estimate each temperature measurement to within 0.01 ℃. Stir the solution continuously at a uniform rate. occasionally stir the ice-salt water bath.

16. Record temperature-time data until the freezing region is reached and until the readings are essentially constant or decreasing only slightly over several minutes.

If time permits, do a second determination using a second portion of the unknown and a fresh distilled water sample.

17. Discard all solutions as directed by your laboratory instructor. Disassemble the apparatus. Wash glassware with detergent, rinse, and drain to dry.

【Data Analysis】

Record all calculated results on Data Sheet 2.

1. Using your temperature-time data from Part I and the coordinate-paper, prepare a cooling curve for pure water by plotting temperature (ordinate) versus time (abscissa). To determine the freezing point of your water sample, extend the horizontal segment of the curve (representing the period during which freezing occurred) until it intersects the left-hand portion of the curve (see Figure 2). Locate the temperature corresponding to this intersection point on the ordinate of the graph.

2. Using your temperature-time data from Part II and the coordinate-paper, prepare a cooling curve for the solution. Use it to determine the freezing point of the solution.

3. Calculate the mass of water, in grams, you used as the solvent in Part II. Express this mass in kilograms. Calculate the mass of unknown, in grams, you used as the solute.

4. Calculate the freezing point depression for the solution (Equation(2 - 10 - 1)).

5. Calculate the molality of the solution (Equation(2 - 10 - 2)).

6. Calculate the molar mass of the unknown (Equation(2 - 10 - 4)).

7. Repeat Calculations 1 ~ 6 using the data from your second determination in each part, if you did them.

8. If you did two determinations in each Part, calculate the average freezing point of water and the average molar mass of the unknown.

【Data Sheet 1】

1. Determining the Freezing Point of Water.

time, min	determination 1 temp, ℃	determination 2 temp, ℃	time, min	determination 1 temp, ℃	determination 2 temp, ℃
0.0	_____	_____	7.5	_____	_____
0.5	_____	_____	8.0	_____	_____
1.0	_____	_____	8.5	_____	_____
1.5	_____	_____	9.0	_____	_____
2.0	_____	_____	9.5	_____	_____
2.5	_____	_____	10.0	_____	_____
3.0	_____	_____	10.5	_____	_____
3.5	_____	_____	11.0	_____	_____
4.0	_____	_____	11.5	_____	_____
4.5	_____	_____	12.0	_____	_____

time, min	determination 1 temp /℃	determination 2 temp /℃
5.0	_____	_____
5.5	_____	_____
6.0	_____	_____
6.5	_____	_____
7.0	_____	_____

time, min	determination 1 temp /℃	determination 2 temp /℃
12.5	_____	_____
13.0	_____	_____
13.5	_____	_____
14.0	_____	_____
14.5	_____	_____

【Data Sheet 2】

2. Determining the Freezing Point of Water.

unknown identification code _____

	determination	
	1	2
mass of test tube, water, and unknown, g	_____	_____
mass of test tube and unknown, g	_____	_____
mass of test tube, g	_____	_____

time, min	determination 1 temp /℃	determination 2 temp /℃
0.0	_____	_____
0.5	_____	_____
1.0	_____	_____
1.5	_____	_____
2.0	_____	_____
2.5	_____	_____
3.0	_____	_____
3.5	_____	_____
4.0	_____	_____
4.5	_____	_____
5.0	_____	_____
5.5	_____	_____
6.0	_____	_____
6.5	_____	_____
7.0	_____	_____

time, min	determination 1 temp /℃	determination 2 temp /℃
7.5	_____	_____
8.0	_____	_____
8.5	_____	_____
9.0	_____	_____
9.5	_____	_____
10.0	_____	_____
10.5	_____	_____
11.0	_____	_____
11.5	_____	_____
12.0	_____	_____
12.5	_____	_____
13.0	_____	_____
13.5	_____	_____
14.0	_____	_____
14.5	_____	_____

	determination 1	2
Mass of unknown, g	_____	_____
Mass of water, g	_____	_____
Mass of water, kg	_____	_____
Freezing point of water, ℃	_____	_____
Average freezing point of water, ℃	_____	_____
Freezing point of unknown solution, ℃	_____	_____
Freezing point depression for solution, Δt_f, ℃	_____	_____
Molality of unknown, g/mol	_____	_____
Molar mass of unknown, g/mol	_____	_____
Average molar mass of unknown, g/mol	_____	

【Post Lab Questions】

1. What effect would each of the following have on the calculated molar mass of an unknown, as determined using the Procedure described in this experiment? Briefly explain.

(a) The balance was not zeroed before the three weighings of the test tube, sample, and water were made. This caused each mass measurement to be 0.102 g greater than it should have been.

(b) The thermometer consistently read 0.06 ℃ lower than it should have, over the entire experimental temperature range.

(c) The masses of the unknown and the solvent (water) were determined, and the unknown completely dissolved in the water. Just before the jacketed test tube containing the unknown solution was placed in the ice-salt water bath, a small amount of the solution was spilled out of the tube.

(d) The unknown contained a small amount of an insoluble impurity.

(e) The unknown contained a small amount of a soluble impurity, which had a molar mass lower than that of the pure unknown.

(f) A portion of the unknown did not dissolve.

2. An unlabeled bottle contains an unknown potassium halide salt, KX. A student must determine the identity of the salt by using freezing point depression data. Because KX is a strong electrolyte, one mole of KX yields two moles of ions in solution. The student finds that when 1.13 g of the unknown salt is dissolved in 24.65 g of distilled water, the freezing point of the resulting solution is −1.01 ℃. The freezing point of the distilled water is determined to be +0.02 ℃.

(a) Calculate the freezing point depression for the solution.

(b) Calculate the molality of the solution.

(c) Calculate the molar mass of the unknown salt.

(d) Given that the salt is a potassium halide, KX, identify X, the halide ion.

第3章 综合性实验

实验11 B-Z振荡反应

【实验目的】

1. 了解B-Z反应的基本原理；
2. 掌握利用微机系统研究化学振荡反应的方法；
3. 通过测定电位-时间曲线求得化学振荡反应的表观活化能；
4. 加深理解振荡反应这一现象，初步认识体系非平衡态下的复杂行为；
5. 通过本实验理解化学热力学、化学动力学和电化学知识的综合运用；
6. 初步理解自然界中普遍存在的非平衡、非线性问题。

【预习要求】

1. 了解B-Z反应的基本原理及B-Z振荡反应系统的特点；
2. 了解B-Z振荡反应仪器使用方法及软件使用方法；
3. 了解本实验的注意事项。

【实验原理】

人们通常所研究的化学反应，其反应物和产物的浓度呈单调变化，最终达到不随时间变化的平衡状态。而某些化学反应体系中，会出现非平衡、非线性现象，即有些反应物和产物的浓度会呈现周期性变化，该现象称为化学振荡。为了纪念最先发现、研究这类反应的两位科学家（Belousov和Zhabotinskii），人们将可呈现化学振荡现象的含溴酸的反应系统笼统地称为B-Z振荡反应（B-Z Oscillating Reaction）。

大量的实验研究表明，化学振荡现象的发生必须满足三个条件：①必须是远离平衡的敞开体系；②反应历程中应含有自催化步骤；③必须具有双稳态性，即可在两个稳态间来回振荡。

有关B-Z振荡反应的机理，目前为人们所普遍接受的是FKN机理，即由Field、Körös和Noyes三位学者提出的机理。下面以$BrO_3^- \sim Ce^{4+} \sim CH_2(COOH)_2 \sim H_2SO_4$体系为例加以说明。该体系的总反应为：

$$2H^+ + 2BrO_3^- + 3CH_2(COOH)_2 \longrightarrow 2BrCH(COOH)_2 + 3CO_2 + 4H_2O$$

$$(3-11-1)$$

体系中存在着下面几个反应过程。

A 过程：

$$BrO_3^- + Br^- + 2H^+ \xrightarrow{k_1} HBrO_2 + HOBr \qquad (3-11-2)$$

$$HBrO_2 + Br^- + H^+ \xrightarrow{k_2} 2HOBr \qquad (3-11-3)$$

B 过程：

$$BrO_3^- + HBrO_2 + H^+ \xrightarrow{k_3} 2BrO_2 + H_2O \qquad (3-11-4)$$

$$BrO_2 + Ce^{3+} + H^+ \xrightarrow{k_4} HBrO_2 + Ce^{4+} \qquad (3-11-5)$$

$$2HBrO_2 \xrightarrow{k_5} BrO_3^- + HOBr + H^+ \qquad (3-11-6)$$

Br^- 的再生 C 过程：

$$4Ce^{4+} + BrCH(COOH)_2 + H_2O + HOBr \xrightarrow{k_6} 2Br^- + 4Ce^{3+} + 3CO_2 + 6H^+ \qquad (3-11-7)$$

当 $[Br^-]$ 足够高时，主要发生 A 过程[反应(3-11-2)、(3-11-3)]，其中第(3-11-2)步是速率控制步，当达到稳态时，有

$$[HBrO_2] = \frac{k_1}{k_2}[BrO_3^-][H^+] \qquad (3-11-8)$$

当 $[Br^-]$ 低时，主要发生 B 过程[反应(3-11-4)、(3-11-5)、(3-11-6)]，Ce^{+3} 被氧化，其中第(3-11-4)步是速率控制步，反应经(3-11-4)、(3-11-5)将自催化产生 $HBrO_2$，达到稳态时有

$$[HBrO_2] \approx \frac{k_3}{2k_5}[BrO_3^-][H^+] \qquad (3-11-9)$$

由反应(3-11-3)和(3-11-4)可以看出：Br^- 和 BrO_3^- 是竞争 $HBrO_2$ 的。当 $k_2[Br^-] > k_3[BrO_3^-]$ 时，自催化过程(3-11-4)不可能发生。自催化是 B-Z 振荡反应中必不可少的步骤，否则该振荡不能发生。

研究表明，Br^- 的临界浓度为

$$[Br^-]_{crit} = \frac{k_3}{k_2}[BrO_3^-] = 5 \times 10^{-6}[BrO_3^-]$$

若已知实验的初始浓度 $[BrO_3^-]$，由式(3-11-10)即可估算 $[Br^-]_{crit}$。

通过反应(3-11-7)可实现 Br^- 的再生。

综上所述，B-Z 振荡反应体系中振荡的控制物种是 Br^-。体系中存在着两个受溴离子浓度控制的过程 A 和 B，当 $[Br^-]$ 高于临界浓度 $[Br^-]_{crit}$ 时，发生 A 过程；当 $[Br^-]$ 低于 $[Br^-]_{crit}$ 时发生 B 过程。也就是说 $[Br^-]$ 起着开关作用，它控制着从 A 到 B 过程的发生，再由 B 到 A 过程的转变。在 A 过程中，由于化学反应，$[Br^-]$ 降低，当 $[Br^-]$ 达到 $[Br^-]_{crit}$ 时，

B 过程发生。在 B 过程中,B 中产生的 Ce^{4+} 通过 C 过程使 Br^- 再生,$[Br^-]$ 增加,当 $[Br^-]$ 达到 $[Br^-]_{crit}$,A 过程发生,从而完成一个循环。

从 FKN 机理可以看出,系统中 $[Br^-]$、$[HBrO_2]$ 和 $[Ce^{4+}]$、$[Ce^{3+}]$ 都随时间做周期性地变化。在实验中,可以采用电化学方法,在不同温度下通过测定 $[Ce^{4+}]$ 和 $[Ce^{3+}]$ 之比产生的电势随时间变化曲线,分别从曲线中求出诱导时间 $t_{诱}$ 和振荡周期 $t_{振}$,并根据阿伦尼乌斯(Arrhenius)方程得

$$\ln(1/t_{诱})(或\ 1/t_{振}) = -\frac{E}{RT} + \ln A$$

分别以 $\ln(1/t_{诱})$ 对 $\frac{1}{T}$ 作图和以 $\ln(1/t_{振})$ 对 $\frac{1}{T}$ 作图,从图中可求出表面活化能 $E_{诱}$ 和振荡周期活化能 $E_{振}$。

综合运用化学热力学、化学动力学和电化学知识是研究化学振荡反应的一般方法。

【仪器与试剂】

仪器:超级恒温水浴 1 台,B-Z 振荡反应数据接口系统 1 套,计算机 1 台,双层玻璃反应器 1 个,铂电极 1 个,参比电极 1 个,磁力搅拌器 1 台。

药品:0.50 mol·L^{-1} 丙二酸(用分析纯配制),3.00 mol·L^{-1} 硫酸,0.25 mol·L^{-1} 溴酸钾(用分析纯配制),0.004 mol·L^{-1} 硫酸铈铵(用分析纯在 0.02 mol·L^{-1} 硫酸介质中配制)。

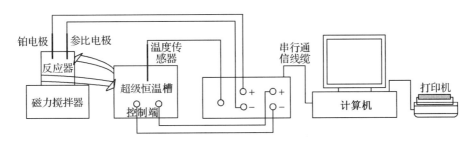

图 3-11-1　B-Z 振荡反应数据采集接口装置

【实验步骤】

1. 按图示检查仪器连线。铂电极接接口装置电压输入正端(+),参比电极接接口装置电压输入负端(-)。将接口装置的温度传感器探头插入恒温水浴中,并固定好。

2. 接通恒温水浴电源,将恒温水通往反应器,将恒温水浴温度设定至 25.0 ℃。接通 B-Z 振荡反应数据采集接口装置电源,启动计算机,运行 B-Z 振荡反应实验软件,进入主菜单。

3. 进入参数设置菜单,设置横坐标极值 800 s,纵坐标极值 1 220 mV;纵坐标零点 800 mV;起波阈值 6 mV;画图起始点设定为实验一开始就画图;目标温度 25.0 ℃,然后确定,退出。

4. 进入开始实验菜单,等待恒温水浴达到设定温度并在软件窗口出现提示,并确认。

5. 取硫酸铈铵溶液 15 mL,放入一锥形瓶中,置于恒温水浴中。在玻璃反应器中放入洁净的磁力搅拌转子,然后依次加入已配好的丙二酸溶液、溴酸钾溶液、硫酸溶液各 15 mL,盖上胶塞,使电极浸入溶液。打开磁力搅拌器,并调节好搅拌速度以溶液旋转但又不起漩涡为宜,实验过程中不得改变搅拌速度。

6. 当系统控温完成出现提示后再恒温 5 min,然后单击"开始实验"键,根据提示,选择"保存实验波形",输入 B-Z 振荡反应即时数据,存储文件使用默认的文件扩展名(.dat),但此时不要单击输入文件名窗口中的"OK"键,全部实验要测定并记录 6 个温度下的实验数据,每个温度下都需要输入相应的数据存储文件名,全部实验数据将存储在 6 个数据文件中,实验操作结束后要对数据文件中存储的数据进行处理。因此所取文件名应采用编号加以区别,以免文件名重名,若重名后面的数据文件将覆盖前面的数据文件。如李姓同学将 6 个温度下的数据文件取名为 li25.dat,li30.dat,…,li50.dat。

7. 将恒温后的硫酸铈铵溶液 15 mL 加入到反应器中,在加入一半溶液时立即单击输入文件名窗口中的"OK"键,系统开始采集记录显示电位信号并绘图,如图 3-11-2 所示。观察反应器中溶液颜色变化和记录的电位曲线。

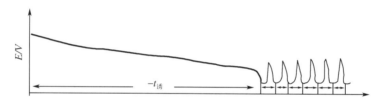

图 3-11-2 诱导期定义图

从加入硫酸铈铵溶液开始画图起始到开始振荡定义为 $t_{诱}$,振荡开始后每个周期依次定义为 $t_1, t_2, t_3, \cdots$。

8. 待画完 10 个振荡周期或曲线运行到横坐标最右端后,单击"停止实验"键,停止信号采集,此时可以单击"查看峰谷值",观察各波的峰、谷值。

9. 点击"打印"按钮,将上图打印出来,每名同学 1 份。(注:此图为振荡反应代表图,只需打印某一个温度的代表图即可,不必每个温度皆打印)

10. 待实验停止后,不要按"退出"按钮,可直接单击"修改目标温度"键,改变温度为 30.0 ℃、35.0 ℃、40.0 ℃、45.0 ℃、50.0 ℃,重复步骤 4~8。观察计算机屏幕上方"信号电压"和"计时",记录下 $t_{诱}$(第一个波峰前的最低信号电压所对应的时间)和 t_9(第五个波谷时的时间)。

11. 实验完成后,单击"退出"键退出,此时会有提示"是否保存实验数据",单击"Yes",出现对话框"请输入保存实验数据文件名",输入文件名后再单击"Yes",将此次实验的不同反应温度下的起波时间保存入文件。

12. 关闭仪器(接口装置、磁力搅拌器、恒温水浴)电源。

【实验记录及数据处理】

1. 根据公式 $t_{振} = \dfrac{t_9 - t_{诱}}{9}$ 求出振荡周期 $t_{振}$。

2. 根据公式 $\ln(1/t_{诱}) = -\dfrac{E_{表}}{RT} + \ln A$($A$,指前参量;$E_{表}$,表观活化能;$T$,绝对温度),作 $\ln(1/t_{诱}) - 1/T$ 图,由斜率求出表观活化能 $E_{诱}$。

3. 作 $\ln(1/t_{振}) - 1/T$ 图,由斜率求出表观活化能 $E_{振}$。

作图有两种方法:进入 B-Z 振荡反应软件,进入"数据处理"菜单,对实验数据进行处理;利用通用数据处理软件完成,如 Origin、Excel 等。

4. 将各个温度下的电位-时间图、$\ln(1/t_{诱}) - 1/T$ 图及 $\ln(1/t_{振}) - 1/T$ 图粘贴到 Word 文档中,写上班号、姓名、实验日期等,在一页纸上完成打印。

5. 对振荡曲线进行解释。

【讨论与说明】

1. 所使用的反应器一定要清洗干净,搅拌子位置及搅拌速度都要加以控制,一般以搅拌时液面形成小漩涡但不产生大量气泡为宜;

2. 小心使用硫酸溶液,避免对实验者和仪器设备造成腐蚀;

3. 根据室温的不同,一般要控制恒温水浴的温度比目标(反应体系)温度高 0.3~0.9 ℃。

【思考题】

1. 影响诱导期的主要因素有哪些?

2. 为什么在实验过程中应尽量使搅拌子的位置和转速保持一致?

3. 什么是化学振荡现象?产生化学振荡需要什么条件?

4. 本实验记录的电势主要代表什么意思?其与能斯特方程求得的电位有何不同?

【相关阅读】

1. B-Z 振荡反应数据采集系统介绍

本实验使用 B-Z 振荡反应数据采集接口系统,并与计算机相连,通过接口系统测定电极(铂电极与参比电极)的电势信号,经通信口传送到计算机,自动采集处理数据,如图 3-11-1 所示。

B-Z 振荡反应数据采集接口系统仪器的前面板上有两个输入通道,用于输入 B-Z 振荡电压信号和温度传感器信号以及一个通断输出控制通道,用于控制恒温水浴。温度传感器用于测温。仪器的后面板上有电源开关、保险丝座和串行口接口插座。具体接线方法:铂电极接电压输入正端(+),参比电极接电压输入负端(-),将仪器后面板上的串行口接计算机的串行口一(必须串行口一)。

B-Z 反应数据采集接口系统软件:双击 Windows 桌面上 B-Z 振荡反应软件图标,即进入软件首页。如果要进入实验,单击继续键进入主菜单。

主菜单:有参数矫正、参数设置、开始实验、数据处理、退出等菜单项。

参数校正菜单:包括两个子菜单项,即"温度参数校正"和"电压参数校正"。电压参数一般情况下不需校正。如需要进行温度参数校正,方法如下:将恒温水浴调至一特定温度,如 20.0 ℃,把温度传感器插入恒温水浴中。进入温度参数矫正子菜单,观察传感器送来的信号,待信号稳定后,输入当前温度值(20.0),单击低点部位的"确定"键。将恒温水浴温度升高到 30.0 ℃,观察传感器的信号稳定后,输入当前温度值(30.0),单击上方的"确定"键,再单击最下方的"确定"键。

参数设置菜单:包括横坐标极值、纵坐标极值、纵坐标零点、起波阈值、目标温度、画图起始点设定;功能按钮包括"确定"和"退出"。

横坐标设置:用于设置绘图区的横坐标范围,单位为 s。

纵坐标极值:用于设置绘图区的纵坐标最大值,单位为 mV。例如,一般 B-Z 振荡实验的电势波动范围为 850~1 100 mV,则可设纵坐标极值为 1 200 mV。

纵坐标零点:用于设置实验绘图区的纵坐标零点,单位为 mV。设置纵坐标极值和零点这两项参数,须根据实验中 B-Z 反应波形的经验值来调整。例如:一般 B-Z 振荡实验的电势波动范围为 850~1 100 mV,则可设纵坐标极值为 850 mV。

起波阈值:当发现起波时间识别不正确时,可以调节起波阈值,可在 1~20 mV 范围内调节,默认设置为 6 mV,一般不需改变。

目标温度:设定实验的反应温度,设定完成后,程序即自动自行控温至目标温度。

画图起始点:设定实验一开始就画图或起波后开始画图。

开始实验菜单:包括开始实验、修改目标温度、查看峰谷值、读入实验波形、打印五个菜单;功能按钮包括"退出"。在此窗口,可采集、记录振荡曲线。

数据处理菜单:包括使用当前实验数据进行数据处理、从数据文件中读取数据和打印三个子菜单;功能按钮包括"退出"。"使用当前实验数据进行数据处理"可将界面上的数据进行处理,或对输入的数据进行处理,画出 $\ln(1/t_{诱})-1/T$ 图并求出表面活化能。

2. 奇妙的化学振荡反应

【实验】配制三种溶液

A. 将 102.5 mL 30% 的双氧水溶液稀释至 250 mL。

B. 10.7 g KIO_3 加 10 mL 2 mol/L 的硫酸,稀释至 250 mL。

C. 取 0.075 g 淀粉溶于少量热水并加 3.9 g 丙二酸和 0.845 g 硫酸锰稀释至 250 mL。

用三支量筒分别量取 A、B、C 各 50 mL,同时加入至洁净的 250 mL 烧杯,用玻璃棒略搅拌后静置。

你认为会出现什么样的现象呢?

让我们暂停一下,来强调这种现象多么出乎意料。假设我们有两种分子,一种是琥珀色,一种是蓝色。由于分子的混乱运动,我们可以想象在给定瞬间有较多琥珀色位于容器某一区域,过一会儿,有较多蓝色聚集等。这样,我们观察到的溶液呈现黑或灰色,有可能偶然而不规则地闪现琥珀色或蓝色。但是,事实并非如此,系统开始完全无色,然后溶液突然变为琥珀色,然后又变为无色(相当短暂),迅速又变为蓝色,之后溶液的颜色就在琥珀色与蓝色之间振荡,并且所有这些改变都以规则的时间间隔发生,维持着一个恒定周期自动变化。

这个反应就是 1960 年初期被发现的"别罗索夫-柴波廷斯基"反应,简称 B-Z 反应。相似的反应模式还有俄冈器、布鲁塞尔器,但它们要更复杂些。这些反应被称为化学钟。

在上面的实验里存在着这样的五个反应:

$$2KIO_3 + 5H_2O_2 + H_2SO_4 = I_2 + K_2SO_4 + 6H_2O + 5O_2\uparrow$$

$$I_2 + 5H_2O_2 + K_2SO_4 = 2KIO_3 + 4H_2O + H_2SO_4$$

$$I_2 + CH_2(COOH)_2 \longrightarrow CHI(COOH)_2 + I^- + H^+$$

$$I_2 + CHI(COOH)_2 \longrightarrow CI_2(COOH)_2 + I^- + H^+$$

$$I^- + I_2 = I_3^-$$

丙二酸的加入是为了以 I_3^- 的形式"贮存" I_2，以增大 I_2 的溶解度。这样能延长变色时间周期和循环次数。有人曾做过加热实验，在加热情况下，产生的 I_2 很快以碘气的形式挥发使反应中断。显然蓝色是由碘分子与淀粉溶液作用的结果。溴和氯的衍生物是可以代替碘化物的，但"显色剂"不同，这时用试亚铁灵（硫酸亚铁二氮杂菲），在适当的催化剂如铈、锰或试亚铁灵的存在下有机酸（丙二酸）被溴酸盐氧化。五个反应累加结果是 $2H_2O_2 \longrightarrow 2H_2O + O_2 \uparrow$。如果向反应器中不断加入溴酸盐、丙二酸、双氧水反应物，同时产物通过溢流管不断离开反应器，这样可以使化学钟无限期走下去。

上述所说的振荡可称之为时间振荡，更令人不可思议的是空间振荡。在培养皿中倒入薄薄的一层 B-Z 反应物，放入不与之反应的颗粒，如沙粒或 1 cm 左右的牙签尖端等。慢慢会产生以这些点为中心的同心圆花纹，以与底色不同的颜色呈环状向外扩展。这个实验用的是溴的衍生物，观察到的图像好像一串规则间隔的白色环带，贯穿于静止的红色混合物中，环带很狭窄，中间呈明显的白色，两缘弥散，移动速度约每秒 1 mm，直至扩散到全部为白色。这种空间的周期性变化称为化学波。

化学钟和化学波很早就被发现过，但它们被早期科学家所厌恶，它们不是科学家所希望的平衡态。对于化学反应是怎样发生的，我们从书上学到的是这样一种看法：在空间中浮动的分子彼此碰撞，当发生"反应碰撞"时，以新的形式再现。这无疑是无序的行为，但化学钟和化学波所有的分子以一定的时间间隔，同时改变了它们的化学性质。在布鲁塞尔器和 B-Z 反应中利用计算机来模拟化学波及溴或碘离子的浓度，这样的坐标图非常漂亮地显示着类似正弦曲线一样的严格的规则性，这是否意味它们分子之间具有一种"通信"手段呢？问题是描述它的参数是宏观的，不是分子间距的 10^{-10} m 的数量级，而是 10^{-2} m 的数量级。时间的尺度也不同，它们不是分子的时间（某种分子具有 10^{-15} s 左右的一种振荡周期），而是数秒、数分钟甚至数小时（前文所述的 A、B、C 实验中每种颜色持续时间约 17 s，溴衍生物则约 2 min，A、B、C 的反应至结束用了几分钟，而溴衍生物则花了好几个小时才结束）。在化学钟和化学波中，系统作为一个整体，体系中的分子自己组织起来形成时空上的一致行动。在生物学中也有催化剂，它们甚至更有效，那是些特殊的蛋白质即酶。我们学过在某些酶的参与下 ATP 与 ADP 可以互相转化。一些生化实验已经发现，与葡萄糖被分解循环过程中有关的 ATP 与 ADP 的浓度方面存在着时间性的振荡，并且发现几种关键酶控制的反应也是在远离平衡的条件下进行的。至于空间方面，则有著名的"阿米巴聚集"。

3. 耗散结构理论

1968年,比利时化学家 Prigogine IR 在历经了近20年的探索以后,提出了耗散结构理论。他指出:一个开放体系在达到远离平衡态的非线性区域时,一旦体系的某一个参量达到一定阈值后,通过涨落就可以使体系发生突变,从无序走向有序,产生化学振荡一类的自组织现象。这里,实质上提出了产生有序结构的以下四个必要条件。

(1) 开放体系:这样才可能同外界交换物质与能量,形成有序结构。具体来说,这样才可能从外界向体系输入反应物等来使体系的自由能或有效能量不断增加,即有序度不断增加;同时,才可能从体系向外界输出生成物等来使体系无效能不断减少,即无序度或熵量不断减少。前者是向体系输入负熵,后者是从体系输出正熵,从而使体系的总熵量增长为零或为负值,以形成或保持有序结构。输入负熵是消耗外界有效物质与能量的过程,输出正熵是发散体系无效物质与能量的过程。这一耗一散就成了产生自组织有序结构的必要条件。因此,自组织有序结构也可以称为耗散结构。显然,耗散结构在非开放体系中是不可能形成或保持的。

(2) 远离平衡态:这样才可能使体系具有足够的反应推动力,推进无序转化为有序,形成耗散结构。如在恒温恒压条件下,可以使反应物浓度远高于平衡浓度,生成物浓度远低于平衡浓度,从而在实际浓度与平衡浓度间造成巨大浓度差,以推进化学振荡反应的产生。相反,如果在平衡态,则实际浓度与平衡浓度相等,二者之差为零,反应推动力为零,反应已经达到极限,反应体系的浓度已经不再随时间变化发生任何变化,即已经达到"时间终点"。因此,也就不可能产生浓度随时间、空间而发生周期性变化的化学振荡现象。此外,在平衡态,体系的熵量已经增至极大,无序度已经增至极大,从而也不可能产生有序。所以Prigogine IR 说,非平衡是有序之源。形象地看,这好比是往咖啡里面加牛奶,达到平衡时的最后状态只能是一碗混沌无序的灰色浑汤。但是在达到那个状态以前的非平衡态,则白牛奶在黑咖啡里排演了多少瞬息万变的漩涡花样和结构!可见,有序的生机是在远离平衡态时萌动的。

(3) 非线性作用:这是一种所得超所望的非线性因果关系,即一个小的输入就能产生巨大而惊人的效果。这样才可能使体系具有自我放大的变化机制,产生突变行为和相干效应、协同动作,以异乎寻常的方式重新组织自己,实现有序。相反,如果只是具有线性作用,要素间的作用只能是线性叠加即量的增长而不能产生质的飞跃和实现有序。

这种非线性作用在化学体系中体现在反应链上存在着自催化或交叉催化的环节,即某些反应物分子的一个生成物正是它们自身所需要的催化剂,从而使反应速率达到雪崩式的加快(自催化);或属于两个不同反应链上的两个产物能各自催化对方的反应(交叉催化),其结果是可以产生一种难以控制的剧变行为。这种自催化或交叉催化产生的剧变行为在技术控制论中被称为正反馈,即某种对于指定参考值的偏差不仅未能消除反而得到加强的行为。在化学振荡反应中,正是由于具有了正反馈,才使体系得以造成失稳、活化、放大成化学钟里的前后呼应的颜色变化,产生周期性的振荡。实际上,正反馈是一种自我复制、自我放大的变化机制,因此才能使亿万分子的微观行为像得到指令般地协同动作并在宏观上实现有序。可见,正反馈、自催化或交叉催化或非线性的相互作用是产生化学耗散结构的不可缺少的动力条件。

(4) 涨落作用:即体系中温度、压力、浓度等某个变量或行为与其平均值发生偏差的作用。体系具有涨落或起伏的变化才能启动非线性的相互作用,使体系离开原来的状态,发生质的变化,跃迁到一个新的、稳定的有序态,形成耗散结构。因此,涨落是一种启动力,涨落导致有序。涨落主要是由于受到体系内部或外部的一些难以控制的复杂因素干扰引起的,具有随机的偶然性,然而却可以导致必然的有序。这就再一次表明,必然性要通过偶然性来表现,偶然性是必然性的补充。

4. "耗散结构"理论跨学科的应用

耗散结构不仅存在于化学领域,而且也普遍存在于整个自然界乃至人类社会的各个领域。因此,耗散结构理论是一种横跨化学学科及整个自然科学和社会科学的理论工具,是一门普遍化热力学或普适性理论,具有广泛的、重要的科学意义。

在化学方面,耗散结构理论除了在化学工业中连续化生产的不平衡体系中得到广泛应用外,还使化学家在理论认识上产生了一个飞跃,即化学自组织反应中与外界进行物质与能量交换的"新陈代谢",也和生物体系一样,是其存在的不可缺少条件,从而使化学体系"活化"了。这就进一步消除了生命与非生命体系的森严壁垒。同时,对于化学中物质的认识也不再是机械论世界观中所描述的那种被动的实体,而是与自发的活性相连的客体。Prigogine IR 认为,"这个转变是如此深远,"甚至可以说是一种"人与自然的新的对话"。所以,现代化学研究已经日益明显地把注意力从平衡态转向非平衡态,从简单的线性关系转向复杂的非线性关系,并成为化学发展的一个重要前沿。耗散结构理论也被誉为是 20 世纪 70 年代化学领域的一项辉煌成就。研究化学耗散结构中亿万分子协同动作的通信手段,则可能为物理学和神经生理学的通信过程找到一种更简单的机制;研究具有完全确定振荡周期的化学钟,则可能研制出比机械振荡的弹簧更加可靠的计时器;研究化学振荡螺旋波与太空星体的旋涡星系、飓风形成的气旋涡和心脏病发作的波动等相似之处,则可能有力地促进天文学、气象学和医学的发展。

在社会领域,由于社会中的各种团体、组织、机构、单位等都可以认为是具有不同层次耗散结构的体系,可以运用耗散结构理论来加以研究。如需要提供良好的开放条件,加强与外界物质、能量和信息的交流以提高体系的有序度,应当保持体系的不平衡态来不断产生新的发展动力,争取发挥整体大于部分之和的非线性放大作用,实现新的飞跃等,从而可以促进整个社会的稳定、有序和进化,形成高度有效的自组织结构。Prigogine IR 认为,社会进化固然有其自身的特点,然而从根本上说也是物理宇宙进化的一个方面。因此,物理、化学上的耗散结构理论也应适于社会进化的研究。《化学科学发展战略》指出,今后进一步开展非平衡热力学的理论与实验研究是一个一旦有所突破会对科学、经济或社会的发展产生重大影响的研究方向。

传统观念的突破、化学耗散结构理论的建立在思想方法上给人以深刻启迪,突破了传统观念,获得了更为全面的科学认识,促进了科学思维方法的发展。

(1) 物理学和生物学规律的统一:过去认为,物理学克劳修斯的热力学第二定律和生物学达尔文的进化论在反映自然规律方面是相互矛盾的。前者认为一个孤立的物理体系总是趋于熵增加的方向,即从有序趋向无序,从高级趋向低级,不断退化;后者认为生物体系居于主导地位的方向总是从无序趋向有序,从低级趋向高级,不断进化。现在耗散结构理论告诉我们,二者并不矛盾,达尔文进化论也符合热力学第二定律。生物体系之所以能从无序趋向有序,根本的原因在于它是一个开放体系,能够不断地从环境向体系输入有效的物质和能量,即负熵流,从而抵消了体系内无效能即正熵量的增加,直至实现有序。这里不仅没有违背热力学第二定律的熵增加原理,相反,是以负熵增加的观点补充、丰富、证明和扩大了它的应用范围,即从孤立系统扩大到了开放系统,从平衡态扩大到了非平衡态,从正熵增加扩大到了负熵增加,从而能够用热力学第二定律的熵增加原理统一揭示物理体系退化和生物体系进化过程的机制与条件,解决了两个规律之间长期以来存在的矛盾。此外,从环境向体系输入负熵,实际也是消耗环境负熵而增大正熵的过程,同时还由于输入和摄取负熵过程中出现的不可避免的热散失而进一步增大了环境的正熵,给环境造成了更大的混乱或无序。这就是说,体系内熵的减少是以环境熵的更大增加为代价取得的。因此,尽管体系内的变化是趋于熵的减少,从无序趋向有序,而就环境和体系的总体变化来说,则仍然趋向于熵增加的方向,即从有序趋向无序,仍然符合热力学第二定律。这样,人们对于热力学理论就可以有更广泛和更全面的理解,并大体上说明了为什么在一个熵递增的环境里,像人类这样具有高度有序结构的生物能够从混乱中出现,从而破除了百年来人们关于热力学第二定律只能破坏有序的传统观念。这是 Prigogine IR 的耗散结构理论作出的重大贡献,为此他获得了 1977 年的诺贝尔化学奖。

(2) 平衡态和非平衡态的并重:过去人们多只侧重于平衡态的研究,诸如对于热平衡、相平衡、电离平衡、化学平衡等平衡规律的研究,似乎只有平衡态才能体现出事物的规律性,而对于非平衡态研究则有所忽视。现在,耗散结构理论告诉我们,非平衡态恰恰正是产生自组织有序结构的一个不可缺少的必要条件,非平衡才是有序之源,必须给予足够重视。此外,宇宙万物种种生动诱人的现象绝大多数都是处于非平衡态而不是平衡态。因此,在重视平衡态研究的同时也重视非平衡态研究,就会更加接近自然界的实际,取得更好效果。总之,耗散结构理论的建立和非平衡热力学的诞生破除了长期以来忽视非平衡研究的传统观念,成为科学界重视非平衡研究之始。

(3) 无序自发向有序的转化:过去认为,从无序到有序是不能自发转化的,否则就违背了热力学第二定律。现在耗散结构理论告诉我们,这种自发转化是可能的,而转化条件实际上也就是依靠开放体系从环境向体系输入的负熵流等形成自组织有序结构的四个条件。它们能把体系内亿万个分子一一准确地安排在特定位置上,并按照确定的时空变化协同动作,发挥作用。这样,耗散结构理论就找到了过去认为不可能存在的从无序自发转化为有序的转化机制与条件,第一次全面掌握了无序和有序之间的双向转化规律。具体说就是在一定条件下,在封闭的平衡体系中将是自发地从有序趋向无序;在开放的非平衡体系中将是自发地从无序趋向有序,从而揭示了无序和有序转化同体系的封闭与开放、平衡与不平衡等条件的联系,建立了更为全面的自然观和科学观,促进了科学和哲学的发展。此外,对于宇宙的未来,依靠远离平衡的开放体系的条件,就可以如恩格斯所预言的那样:放射到宇宙空间中去的热,能够重新集结和活动起来。从无序趋向有序,使体系重新得到"活性"。这就进一步批判了克劳修斯从热力学第二定律片面外推导出的热寂说,即宇宙不会导致完全热静止或完全无序,从而有力捍卫了辩证唯物主义的自然观。

Experiment 11 B-Z Oscillation Reaction

【Purpose of the Experiment】

1. Understand the basic principles of Belosov-Zhabotinskii shock response;

2. Master the method of using a microcomputer system to study chemical oscillation reactions;

3. Obtain the apparent activation energy of the chemical oscillation reaction by measuring the potential-time curve;

4. Deepen the understanding of the phenomenon of oscillating reaction, and initially understand the complex behavior of the system under non-equilibrium state;

5. Understand the comprehensive application of chemical thermodynamics, chemical kinetics and electrochemistry through this experiment;

6. A preliminary understanding of non-equilibrium and nonlinear problems that are common in nature.

【Preview Requirements】

1. Understand the basic principles of the B-Z reaction and the characteristics of the B-Z oscillatory reaction system;

2. Understand how to use the B-Z oscillatory reaction instrument and how to use the software;

3. Understand the precautions of this experiment.

【Principle】

In the chemical reaction that people usually study, the concentration of reactant and product changes monotonously, and finally reaches an equilibrium state that does not change with time. In some chemical reaction systems, non-equilibrium and nonlinear phenomena will appear, that is, some concentrations will show periodic changes. This phenomenon is called chemical oscillation. In order to commemorate the two scientists (Belosov and Zhabotinskii) who first discovered and studied this type of reaction, the reaction system containing bromic acid that can exhibit chemical oscillations is generally called the B-Z Oscillating Reaction (B-Z Oscillating Reaction).

A large number of experimental studies have shown that the occurrence of chemical oscillations must meet three conditions: (1) an open system far from equilibrium; (2) an autocatalytic step should be included in the reaction process; (3) it must be bistable. In this way, it can oscillate back and forth between the two steady states.

Regarding the mechanism of the B-Z oscillation reaction, the FKN mechanism is currently generally accepted by people. Three scholars containing Field, Körös and Noyes proposed this mechanism. Next, take the BrO_3^- ~ Ce^{4+} ~ $CH_2(COOH)_2$ ~ H_2SO_4 system as an example. The overall reaction of the system is:

$$2H^+ + 2BrO_3^- + 3CH_2(COOH)_2 \longrightarrow 2BrCH(COOH)_2 + 3CO_2 + 4H_2O$$
$$(3-11-1)$$

The following reaction process exists in the system.

Process A:
$$2H^+ + BrO_3^- + Br^- \xrightarrow{k_1} 2HBrO_2 + HOBr \qquad (3-11-2)$$
$$H^+ + HBrO_2 + Br^- \xrightarrow{k_2} 2HOBr \qquad (3-11-3)$$

Process B:
$$H^+ + HBrO_2 + BrO_3^- \xrightarrow{k_3} 2BrO_2 + H_2O \qquad (3-11-4)$$
$$H^+ + BrO_2 + Ce^{3+} \xrightarrow{k_4} HBrO_2 + Ce^{4+} \qquad (3-11-5)$$
$$2HBrO_2 \xrightarrow{k_5} BrO_3^- + HOBr + H^+ \qquad (3-11-6)$$

The regeneration process C of Br^-:
$$4Ce^{4+} + BrCH(COOH)_2 + H_2O + HOBr \xrightarrow{k_6} 2Br^- + 4Ce^{3+} + 3CO_2 + 6H^+$$
$$(3-11-7)$$

When Br^- is high enough, process A (reactions $(3-11-2)$, $(3-11-3)$) mainly occur. The step $(3-11-2)$ is the rate control step. When reaching a steady state, there are:

$$[HBrO_2] = \frac{k_1}{k_2}[BrO_3^-][H^+]$$

When $[Br^-]$ is low, process B (reactions $(3-11-4)$, $(3-11-5)$, $(3-11-6)$) mainly occur. Ce^{3+} is oxidized, and the step $(3-11-4)$ is the rate control step. The reaction will auto catalyze to produce $HBrO_2$ after $(3-11-4)$ and $(3-11-5)$. When it reaches a steady state, there are:

$$[HBrO_2] \approx 2\frac{k_3}{k_5}[BrO_3^-][H^+]$$

It can be seen from reactions $(3-11-3)$ and $(3-11-4)$ that Br^- and BrO_3^- compete for $HBrO_2$. When $k_2[Br^-] > k_3[BrO_3^-]$, the autocatalytic process $(3-11-4)$ cannot occur. Autocatalysis is an indispensable step in the B-Z oscillation reaction, otherwise the oscillation cannot occur.

Studies have shown that the critical concentration of Br^- ($[Br^-]_{crit}$) is:

$$[Br^-]_{crit} = \frac{k_3}{k_2}[BrO_3^-] = 5 \times 10^{-6}[BrO_3^-]$$

If the initial concentration of $[BrO_3^-]$ is known, $[Br^-]_{crit}$ can be estimated from the above formula.

The regeneration of Br^- is achieved through reaction $(3-11-7)$.

In summary, the controlling species in the B-Z oscillating reaction system is Br^-. There are two processes A and B controlled by the concentration of bromide ion in the system. When $[Br^-]$ is higher than $[Br^-]_{crit}$, process A occurs. Process B occurs when $[Br^-]$ is lower than $[Br^-]_{crit}$.

That is to say, [Br⁻] plays the role of a switch. It controls the occurrence of the process from A to B, and then the transition from process B to A. In the process A, due to the chemical reaction, [Br⁻] decreases. When [Br⁻] reaches [Br⁻]$_{crit}$, process B occurs. Ce^{4+} produced in process B is reduced in process C, and [Br⁻] is regenerated at the same time. As [Br⁻] increases to reach [Br⁻]$_{crit}$, process A occurs. So as to complete a cycle.

It can be seen from the FKN mechanism that [Br⁻], [HBrO] and [Ce^{4+}], [Ce^{3+}] all change periodically over time. In the experiment, an electrochemical method can be used to measure the potential ~ time curves produced by the changes in the ratio of [Ce^{4+}] and [Ce^{3+}] at different temperatures. The induction time (t_{IT}) and the oscillation period (t_{OP}) can be calculated from the curve. According to the Arrhenius equation:

$$\ln(1/t_{IT})(\text{or} 1/t_{OP}) = -\frac{E}{RT} + \ln A \qquad (3-11-8)$$

Plot $\ln(1/t_{IT}) \sim 1/T$ and $\ln(1/t_{OP}) \sim 1/T$ to obtain the surface induced activation energy E_{IT} and the activation energy of the oscillation period E_{OP}, respectively.

The comprehensive application of chemical thermodynamics, chemical kinetics and electrochemistry is a general method for studying chemical oscillation reactions.

【Apparatus and Reagents】

Instrument: Super constant temperature bath; B-Z oscillation reaction data interface system, computer, double glass reaction device, platinum electrode, calomel electrode, magnetic stirrer.

Reagents: 0.50 mol·L⁻¹ malonic acid, 3.00 mol·L⁻¹ sulfuric acid, 0.25 mol·L⁻¹ potassium bromate, 0.004 mol·L⁻¹ ammonium cerium sulfate (containing 0.02 mol·L⁻¹ sulfuric acid).

【Procedures】

1. Connect the wiring as shown in Figure 3-11-1. The platinum electrode is connected to the voltage input to the positive terminal (+), and the reference electrode is connected to the voltage input to the negative terminal (−). Insert the temperature sensor probe into the constant temperature water bath and fix it.

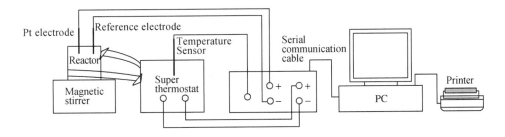

Figure 3-11-1   B-Z oscillation reaction data acquisition interface device

2. Turn on the power of the thermostatic bath, pass the thermostatic water to the reactor, and set the temperature of the thermostat to 25.0 ℃. Turn on the power of the B-Z oscillation reaction data acquisition interface device. Start the computer, run the B-Z oscillation reaction experiment software, and enter the main menu.

3. Enter the parameter setting menu. Set the abscissa extreme value of 800 s, the ordinate extreme value of 1220 mV; the ordinate zero point of 800 mV; the wave threshold of 6 mV; plotting the starting point of the drawing with the beginning of the experiment; the target temperature of 25.0 ℃, then confirm and exit.

4. Enter the start experiment menu. Wait for the thermostat to reach the set temperature and a prompt appears in the software window, and confirm.

5. Take 15 mL of ammonium sulfate solution, put it in an Erlenmeyer flask, and place it in a thermostat water bath. Put a clean magnetic stirring rotor in the glass reactor, and then add 15 mL of the prepared malonic acid solution, potassium bromate solution, and sulfuric acid solution, respectively, and cover with rubber plugs to immerse the electrodes in the solution. Turn on the magnetic stirrer, and adjust the stirring speed so that the solution rotates without vortexing. The stirring speed should not be changed during the experiment.

6. When the system temperature control is completed, click the "Start experiment" button, follow the prompts, select "Save experimental waveform", enter the B-Z oscillation response instant data storage file name, use the default file extension (.dat), but do not click the input file at this time "OK" button in the name window. All experiments need to record the experimental data at 6 different temperatures. Each temperature needs to enter the corresponding data storage file name, all the experimental data will be stored in 6 data files (* .dat), the data stored in the data file should be processed after the experiment operation is over. Therefore, the file names should be distinguished by numbers to avoid duplication of file names. The following data files will overwrite the previous data files. Classmate Li named the data files at 6 temperatures as li25.dat, li30.dat,..., li50.dat.

7. Add 15 mL of cerium ammonium sulfate solution at a constant temperature to the reactor. When half of the solution is added, click the "OK" button in the input file name window, and the system will start to collect, record, display, and draw potential signals, as shown in Figure 3-11-2. Observe the color change of the solution in the reactor and the recorded potential curve.

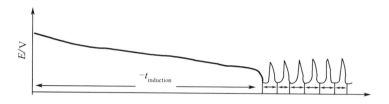

Figure 3-11-2　Definition chart of induction period

From the beginning of adding the cerium ammonium sulfate solution to the beginning of the drawing to the beginning of the oscillation, it is defined as tIT. After the oscillation starts, each period is defined as $t1, t2, t3, \ldots$.

8. After drawing 10 oscillation cycles or the curve runs to the right end of the abscissa, click the "Stop Experiment" button to stop the signal acquisition. At this time, you can click "View Peak and Valley Values" to observe the peak and valley values of each wave.

9. Click the "Print" button to print out the above picture, 1 copy for each student. (Note: This picture is a representative diagram of the oscillation reaction. You only need to print a representative diagram of a certain temperature, not every temperature.)

10. After the experiment is stopped, do not press "Exit", you can directly click the "Modify Target Temperature" button, change the temperature to 30.0 ℃, 35.0 ℃, 40.0 ℃, 45.0 ℃, 50.0 ℃, repeat steps 4~8. Observe the "signal voltage" and "timing" at the top of the computer screen, and record t_{IT} (the time corresponding to the lowest signal voltage before the first peak) and t_9 (the time at the fifth trough).

11. After the experiment is completed, click the "Exit" button to exit. At this time, there will be a prompt: "Whether to save the experimental data", click "Yes". "Please enter the name of the experimental data file", enter "Yes" after the name to save the wave-up time at different reaction temperatures into a file.

12. Turn off the power to the instrument.

【Notes】

1. The reactor used must be cleaned, and the position of the stirrer and the stirring speed must be controlled. Generally, it is advisable to form a small vortex on the liquid surface during stirring but not to produce a lot of bubbles;

2. Use sulfuric acid solution carefully to avoid corrosion to the experimenter and equipment;

3. According to the difference of room temperature, the temperature of the constant temperature bath is generally controlled to be within the range of 0.3~0.9 ℃ higher than the target (reaction system) temperature.

【Data Analysis】

1. Find the oscillation period according to formula $t_{OP} = \dfrac{t_9 - t_{OP}}{9}$.

2. Plot $\ln(1/t_{IT}) \sim (1/T)$ according to formula 3-11-1, and then obtain the apparent activation energy E_{IT} from the slope.

3. Make an $\ln(1/t_{OP}) \sim (1/T)$ graph, and obtain the apparent activation energy E_{OP} from the slope.

4. Paste the potential-time diagram, $\ln(1/t_{IT}) \sim (1/T)$ diagram and $\ln(1/t_{OP}) \sim (1/T)$ diagram at each temperature into a word document, and write the class, name, and experiment date, and then print it out by A4 paper.

5. Explain the oscillation curve.

【Post Lab Questions】

1. What are the main factors affecting the induction period?
2. Why should the position and speed of the stirrer be kept consistent during the experiment?
3. What is a chemical oscillation phenomenon? What conditions are needed to produce a chemical oscillation?
4. What does the electric potential recorded in this experiment mean? What is the difference between the electric potential obtained by the Nernst equation?

实验 12　铝阳极氧化膜电解着色

【实验目的】

1. 了解铝着色的应用；
2. 了解铝阳极氧化膜电解着色的基本原理；
3. 掌握铝阳极氧化电解着色方法；
4. 验证不同电解质体系得到不同的着色颜色。

【预习要求】

1. 了解铝着色的方法；
2. 了解电解着色的原理；
3. 了解本实验的注意事项。

【实验原理】

铝及其合金着色膜具有良好的耐磨、耐晒、耐热和耐蚀性,广泛应用于现代建筑铝型材的装饰防蚀。随着人们生活水平的提高,对家居装饰的要求也越来越高,不仅要求有良好的性能,也需要有亮丽的色彩。

铝阳极氧化膜主要通过有机和电解着色的方法得到所需要的颜色。有机着色是阳极氧化细孔表面物理吸附有机染料而成色,所以是染料本身的颜色,因而有机着色虽可以形成各种鲜艳的颜色,但其耐光、耐候性差,仅用于室内作装饰;而电解着色则是金属元素或氧化物或金属盐在细孔底部近阻挡层析出或还原。电解着色法由于是在铝阳极氧化膜底部沉积一层金属,因而其耐晒、耐蚀、耐磨等性能都较好,但缺点是颜色品种单一,不能从一种颜色跃到另一种颜色,但通过二次氧化,调整细孔结构则能形成"一液电解多色膜",称为电多色,即在同一电解着色液中形成多种颜色。

电解着色的反应本质是金属离子的还原。离子还原析出发生在由阻挡层和多孔层构成的阳极氧化膜微孔底部。要达到上述目的,必须采用交流电解着色。当阳极氧化过的工件上通过交流电时,阻挡层起电容器作用,使交流电在阻挡层中积累。另外,阻挡层的半导体性质起整流作用,这样就把交流电流变成了电容电流(非法拉第电流)和电化学反应电流(法拉第电流)两部分,通过的交流电波形发生变化,工件上负的成分多,金属离子在强的还原和弱的氧化交替作用下被还原析出。随电解着色时间的长短,颜色的变化仅是深浅的变化。

电解着色膜的显色机理是因为膜中沉积的金属胶态粒子对光的反射、吸收、衍射、散射的结果,即胶体分散学说,这种理论是日本佐藤敏彦提出来的,已得到广大研究者的认同。

电解着色是近年来表面工程技术的研究热点,现代工业对于铝型材表面处理提出了更高的要求。电解着色的表面更加均匀一致,没有缺陷和条纹,突破了单调的银白色和古铜色限制,使用性能更加优越,产品质量更加稳定。在我国,近年来铝及铝合金电解着色工艺取得了迅速的发展,已经成为铝及铝合金防腐和装饰的一种独具特色的方法。

【仪器与试剂】

仪器:恒流仪1台,稳压电源1台,恒温水浴1套,100 mL烧杯2个,铁架台1个。
药品:金相砂纸、铝片两片、氢氧化钠、硝酸、硫酸、硫酸亚铁、抗坏血酸、硼酸、高锰酸钾、石墨板。

【实验步骤】

1. 实验装置如图3-12-1所示。

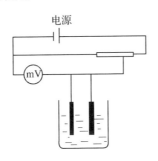

图3-12-1 阳极氧化或着色实验装置

2. 铝阳极氧化膜的制备

取纯度为99%以上、面积为3 cm×3 cm×0.2 cm的两片铝片经如下工艺进行处理,其中电解中阴极和阳极均为铝片。

(1)用金相砂纸对铝表面进行机械抛光,用蒸馏水清洗干净。

(2)把经(1)处理的铝片放入浓度为2.8 mol/L、温度为60 ℃的NaOH溶液中刻蚀30 s,取出用蒸馏水清洗干净。

(3)把经(2)处理的铝片放入室温下浓度为4 mol/L的硝酸溶液中光亮处理20 s,然后取出用蒸馏水清洗干净。

(4)把经(3)处理的铝片放入1.7 mol/L硫酸溶液中阳极氧化20 min,阳极氧化条件为恒压15~17 V,温度为16±1 ℃,取出用蒸馏水清洗干净备用。

3. 着色处理

实验中按照工艺参数配置相应的物质浓度,按电解着色条件进行电解着色。

4. 把步骤 3 中电解着色的铝片用蒸馏水清洗干净,然后烘干观察。

【实验记录及数据处理】

观察着色后铝片的颜色,把结果填入表 3 – 12 – 1 中。

表 3 – 12 – 1 着色条件

实验编号	1		2	
电解液	试剂	浓度/mol·L^{-1}	试剂	浓度/mol·L^{-1}
	$FeSO_4 \cdot 7H_2O$	0.14	$KMnO_4$	0.13
	H_3BO_3	4.85		
	$C_6H_8O_6$(抗坏血酸)	5.70×10^{-3}		
电解条件	电解温度 25~35 ℃ 电解电压 20 V 电解时间 5~10 min		电解温度 20 ℃ 电解电压 15 V 电解时间 5 min	
着色后铝片的颜色				

【讨论与说明】

1. 碱性溶液对铝片表面的腐蚀,可以除去铝材表面的自然氧化膜层;而化学抛光起到整平作用,使其表面光滑没有凹凸;在硫酸溶液中进行阳极氧化,使其表面生成一层多孔的氧化膜,以便进行着色。

2. 在着色过程中不同条件对着色效果有不同的影响。其中电解液浓度只能改变颜色的深浅,这主要是浓度增加时,单位时间内进入氧化膜的显色体增多而引起的;电解着色时间也只影响颜色的深浅。而温度对着色效果的影响最显著,在常温时颜色很淡,到了 30~40 ℃ 时,颜色加深很多,温度是影响扩散速度的最主要原因。

3. 使用强酸强碱要注意安全,用电过程要注意安全。

【思考题】

1. 铝表面没有氧化膜能被电解着色吗?
2. 用直流电能否使铝表面着色?
3. 哪些因素影响着色?还能在铝表面电解着上其他颜色吗?

【相关阅读】

1. 铝阳极氧化膜的着色方法

(1) 整体着色膜。电解整体着色又分为自然发色、电解发色和电源发色,其中电解发色占主导,自然发色次之,电源发色正在开发中。自然发色指阳极氧化过程使铝合金中添加成分(Si、Fe、Mn 等)氧化,而发生氧化膜的着色,如 Al – Si 合金的硫酸阳极氧化膜;电解发色指电解液组成及电解条件的变化而引起氧化膜的着色,如在添加有机酸或无机盐的电解液中阳极氧化,其代表性的技术有 Kalcolor 法(硫酸 + 磺基水杨酸)及 Duranodic 法(硫酸 + 邻苯二甲水杨酸)。

(2) 染色膜。以硫酸一次电解的透明阳极氧化膜为基础,利用氧化膜层的多孔性与化学活性吸附各种色素而使氧化膜着色,根据着色机理和工艺可分为有机染料着色、无机染料着色、色浆印色、套色染色和消色染色等。

(3) 电解着色膜。以硫酸一次电解的透明阳极氧化为基础,在含金属盐的溶液中用直流或交流进行电解着色的氧化膜。1936 年意大利人 Caboni 最早提出了阳极氧化膜的电解着色专利技术,德国人 Elssner 进一步改进了这个方法。20 世纪 60 年代,浅田太平改进并注册了电解着色专利。该专利的特征是,利用交流电为电源,着色溶液采用 Co、Ni、Cu、Ag、Se 的盐类以及它们的含氧盐作为主成分。浅田已经明确鉴别出电解着色工艺过程的几个阶段,包括金属离子进入阳极氧化膜的微孔中,由于电解还原转化成着色的物质等。代表性的电解着色的工业化技术有浅田法(Ni 盐交流着色法)、Anolok 法和 Sallox 法(二者均系 Sn 盐交流着色法)、住化法和尤尼可尔法(Ni 盐"直流"着色和"直流"脉冲着色法)等。20 世纪末在工业上开始得到推广的多色化技术可以在一个电解着色槽中得到多种颜色。这是一种新型的利用干涉光效应的电解着色方法,由于要在电解着色之前增加电解调整,在日本又被称为三次电解法。电解着色膜的耐候性、耐光性和使用寿命比染色膜要好,其能耗与着色成本又远低于整体着色膜,目前已经被广泛用于建筑铝型材的着色。

2. 电解着色膜与染色的区别

(1) 电解着色膜以硫酸一次电解的透明阳极氧化膜为基础,在含金属盐的溶液中用交流进行电解着色的氧化膜(也叫二次电解膜),其电解着色膜的耐候性、耐光性及使用寿命比染色膜好得多,其能耗与着色成本又远低于整体着色膜。目前,电解着色膜广泛用于建筑铝型材着色,但电解着色色调单调,通常仅有古铜色、黑色、金黄色、枣红色等几种,且操作不易控制。

(2) 有机染色。有机染色基于物质的吸附理论。吸附有物理吸附及化学吸附之分,物理吸附是分子或离子以静电力方式的吸附;化学吸附是以化学力方式的吸附,这两者结合起来产生有机染色,通常在一定温度下进行。由于阳极氧化膜孔隙率高,吸附能力强,容易有机染色,这种方法上色快、色泽鲜艳、操作简便,染色后经封孔处理,染料能牢固地附着在膜孔中,提高了膜层的防蚀能力、抗污能力,可以保持美丽的色泽,适用于那些不需要户外使用的大量铝制日用品,室内用铝制工业品以及装饰品等,外观色彩缤纷多样,可满足现代社会人们的审美需求,提高了产品市场的竞争能力。有机染色对氧化膜有一些具体要求:氧化膜层要具有足够的孔隙率;膜孔内壁保持一定的活性;铝在硫酸溶液中得到的阳极氧化膜无色而多孔,因此最适宜染色;氧化膜层必须有一定的厚度,较薄的膜层只能染上很浅的颜色;硬质阳极氧化膜如铬酸常规氧化膜层,均不适合有机染色;氧化膜层应完整、均匀、不应有划伤、砂眼、点腐蚀等缺陷;膜层本身具有合适的颜色,且没有金相结构的差别,如晶粒大小不一或严重偏析等,因此对铝合金材料也有一定的要求,合金成分中硅、镁、锰、铁、铜、铬等含量过高时,往往会引起氧化膜暗哑,则在染色时会产生色调变化。

3. 不同槽液对电解着色膜的影响

国内外工业化的电解着色槽液主要是镍盐或锡盐溶液两大类,其着色膜的颜色基本上都是从浅到深的古铜色系,这是在可见光范围内散射效应得到的色系。实验室里能够电解着色的金属盐很多,不同的金属盐可以得到多种多样不同的颜色,如表 3-12-2 所列为普通硫酸阳极氧化膜在不同金属盐的电解槽液中的着色结果,由于着色膜的性能以及着色槽液的成本与稳定性等原因,多数都不能够产业化和商业化。

表 3-12-2　普通硫酸阳极氧化膜在不同金属盐的电解槽液中的着色结果

电解着色的 金属盐类	电解着色阳极氧化膜的颜色	电解着色的金属盐类	电解着色阳极 氧化膜的颜色
Ni 盐	黄色、青铜色、黑色	Se 盐	红色
Co 盐	黄色、青铜色、黑色	Cr 盐	绿色
Cu 盐	茶色、青铜色、红褐色、黑色	Ba 盐、Ca 盐	不透明白色
Ca 盐、Zn 盐	青铜色系	Mo 盐、W 盐	黄色、蓝色
Ag 盐	绿色	$H_2SO_4 + CuSO_4$	绿色
Au 盐	紫色	$H_3PO_4 + NiSO_4$	绿色、蓝色、红色
$KMnO_4$	金黄色、浅青铜色	氰化亚铁	蓝色

4. 功能性电解着色膜

电解着色膜除了常见的装饰及防腐外,在光学及光电元件、抗菌等方面也有广泛的应用。如在多孔膜内引入 Tb^{3+} 制得的功能化膜,在外加电场的作用下发出绿色光。这种功能化多孔膜能获得较高的发光强度,而且由于多孔膜的孔径极为细小,更可进一步开发出超微细发光元件。抗菌性氧化铝膜是将抗菌成分浸透到膜的孔中并在膜孔中析出,从而使之具有抗菌性作用。如把银、铜、锌等具有抗菌作用的金属离子填充到多孔膜内,然后进行封孔处理就能具有抗菌性。

Experiment 12　Electrolytic Coloring of Aluminum Anodic Oxide Film

【Purpose of the Experiment】

1. Understand the application of aluminum coloring;
2. Understand the basic principles of electrolytic coloring of aluminum anodic oxide film;
3. Master the electrolytic coloring method of aluminum anodizing;
4. Verify that different electrolyte systems get different colored colors.

【Preview Requirements】

1. Understand the method of aluminum coloring;
2. Understand the principle of electrolytic coloring;
3. Understand the precautions of this experiment.

【Principle】

Aluminum and its alloy colored films have good wear resistance, light resistance, heat resistance and corrosion resistance, and are widely used in the decoration and corrosion resistance of modern architectural aluminum profiles. For example, with the improvement of people's living standards, the requirements for home decoration are getting higher and higher, not only requiring good performance, but also bright colors.

The aluminum anodic oxide film is mainly used for organic and electrolytic coloring to obtain the desired color. Organic coloring is the color formed by the physical adsorption of organic dyes on the surface of anodized pores, so it is the color of the dye itself. Therefore, although organic coloring can form a variety of bright colors, it has poor light resistance and weather resistance and is only used for indoor decoration; and electrolysis coloring is the chromatographic or reduction of metal elements or oxides or metal salts near the bottom of the pores. Since the electrolytic coloring method deposits a layer of metal on the bottom of the aluminum anodic oxide film, its properties such as light resistance, corrosion resistance, and wear resistance are good, but the disadvantage is that the color variety is single, and it cannot jump from one color to another. However, through secondary oxidation and adjusting the pore structure, a "one-liquid electrolytic multicolor film" can be formed, which is called electric multicolor, that is, multiple colors are formed in the same electrolytic coloring solution.

The essence of electrolytic coloring is the reduction of metal ions. Ion reduction and precipitation occur at the bottom of the pores of the anodic oxide film composed of the barrier layer and the porous layer. However, to achieve the above-mentioned purpose, AC electrolytic coloring must be used. When alternating current passes through the anodized workpiece, the barrier layer acts as a capacitor, allowing the alternating current to accumulate in the barrier layer. In addition, the semiconductor properties of the barrier layer play a rectifying role, so that the alternating current becomes a capacitive current (Faraday current) and an electrochemical reaction current (Faraday current). The waveform of the passed alternating current changes, and the negative component on the workpiece is changed. Many metal ions are reduced and precipitated under the alternating action of strong reduction and weak oxidation. With the length of electrolytic coloring time, the change of color is only a change of shade.

The color development mechanism of the electrolytic colored film is the result of the reflection, absorption, diffraction, and scattering of light by the metal colloidal particles deposited in the film, that is, the colloidal dispersion theory. This theory was proposed by Toshihiko Sato, Japan, and has been extensively studied. The identity of the person.

Electrolytic coloring is a research hotspot in surface engineering technology in recent years, and modern industry has put forward higher requirements for surface treatment of aluminum profiles. The surface is more uniform, without defects and streaks, breaking through the monotonous silver-white and bronze limitations, superior performance, and more stable product quality. In our country, the electrolytic coloring process of aluminum and aluminum alloy has achieved rapid development in recent years, and it has become a unique method of aluminum and aluminum alloy anti-corrosion and decoration. This experiment is aimed at the electrolytic coloring of aluminum anodizing.

【Apparatus and Reagents】

Instruments: Constant current meter, stabilized power supply, constant temperature water bath, two 100 mL beakers, iron stand.

Reagents: metallographic sandpaper, two aluminum sheets, sodium hydroxide, nitric acid, sulfuric acid, ferrous sulfate, ascorbic acid, boric acid, potassium permanganate, graphite board.

【Procedures】

1. The experimental device is shown in Figure 3-12-1.
2. Preparation of aluminum anodic oxide film

Two aluminum sheets with a purity of more than 99% and an area of 3 cm × 3 cm × 0.2 cm are pretreated by the following process. In the electrolysis, both the cathode and the anode are aluminum sheets.

A. Removing oxide: use metallographic sandpaper to mechanically polish the aluminum surface and clean it with distilled water.

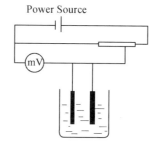

Figure 3-12-1 Experimental apparatus for anodizing or coloring

B. Removing grease: put the aluminum sheet treated by (A) into a NaOH solution with a concentration of 2.8 mol/L at 60 ℃ for 30 s, then take it out and clean it with distilled water.

C. Bright treatment: put the aluminum sheet treated with (B) into a nitric acid solution with a concentration of 4 mol/L at room temperature for 20 s, then take it out and wash it with distilled water.

D. Anodizing: the aluminum plate treated by (C) was anodized in 1.7 mol/L sulfuric acid solution at a constant potential of 15~17 V at 16 ± 1 ℃ for 20 min. Take it out and wash it with distilled water.

3. Coloring treatment

Two coloring process parameters are listed in Table 3-12-1. In the experiment, the corresponding substance concentration was prepared according to the process parameters, and the electrolytic coloring was carried out according to the electrolytic coloring conditions.

4. Wash the electrolytic colored aluminum sheet with distilled water, and then dry it for

observation.

【Notes】

1. The corrosion of the surface of the aluminum sheet by the alkaline solution can remove the natural oxide film layer on the surface of the aluminum material; while the chemical polishing has a leveling effect to make the surface smooth without unevenness; anodize in a sulfuric acid solution to make the surface porous oxide film is formed for coloring.

2. In the coloring process, different conditions have different effects on the coloring effect. Among them, the electrolyte concentration can only change the color depth, which is mainly caused by the increase of color bodies entering the oxide film per unit time when the concentration increases; the electrolytic coloring time also only affects the color depth. The temperature has the most significant effect on the coloring effect. The color is very light at room temperature. When it reaches 30 ~ 40 ℃, the color deepens a lot, because temperature is the most important factor affecting the diffusion rate.

3. Pay attention to safety when using strong acids and bases.

【Data Analysis】

Observe the color of the colored aluminum sheet, and fill in the results in Table 3 – 12 – 1.

Table 3 – 12 – 1 Coloring conditions

Experiment number	1		2	
Electrolyte	Reagent	Concentration /mol·L^{-1}	Reagent	Concentration /mol·L^{-1}
	$FeSO_4 \cdot 7H_2O$ H_3BO_3 Ascorbic acid	0.14 4.85 5.70×10^{-3}	$KMnO_4$	0.13
Electrolysis conditions	Electrolysis temperature: 25 ~ 35 ℃ Electrolysis voltage: 20 V Electrolysis time: 5 ~ 10 min		Electrolysis temperature: 20 ℃ Electrolysis voltage: 15 V Electrolysis time: 5 min	
Color of aluminum sheet after tinting				

【Post Lab Questions】

1. Can the aluminum surface be electrolytically colored without an oxide film?

2. Can the aluminum surface be colored with direct current?

3. What factors affect the coloring? Can other colors be electrolyzed on the aluminum surface?

实验 13　溶胶和乳状液的制备及其性质

【实验目的】

1. 用化学凝聚法制备 $Fe(OH)_3$ 溶胶并测定其 ξ 电势;
2. 验证电解质对溶胶聚沉作用的实验规律;
3. 制备一种乳状液,鉴别其类型并测定其 ξ 电势。

【预习要求】

1. 了解制备溶胶的方法;
2. 了解电泳测定方法;
3. 了解 ξ 电势的意义。

【实验原理】

1. 化学凝聚法制备溶胶

固体以胶体状态分散在液体介质中即称为胶体溶液或溶胶,胶粒直径为 1～100 nm,它是多相系统,有很大的相界面,是热力学不稳定系统。为了形成这种系统并能相对稳定地存在,在制备过程中除了分散相及分散介质外,还必须有稳定剂存在,这种稳定剂可以是外加的第三种物质,也可以是系统内已有的物质。

溶胶的制备方法可分为两大类:(1) 分散法。把较大的物质颗粒变为胶体大小的质点。其常用的分散法有研磨法、电弧法、超声波法、溶胶法。(2) 凝聚法。把物质的分子或离子聚合成胶体大小的质点。凝聚法中的化学凝聚法是一种较为简便的方法,若化学反应生成难溶化合物,那么在一定条件下,就能将此化合物制成溶胶。一般而言,先令化学反应在稀溶液中进行,其目的是使晶粒的增长速度变慢,此时得到的是细小的粒子,即粒子直径为 1～100 nm,使粒子的沉降稳定性得到保证。其次,让一种反应物过量(或反应物本身进行水解的产物),使其在胶粒表面形成双电层,以阻止胶粒的聚集。

如 $FeCl_3$ 在水中即可水解生成红棕色的 $Fe(OH)_3$ 溶胶,其反应为

$$Fe(OH)_3 + HCl \longrightarrow FeOCl + 2H_2O$$

其胶团结构为

$$\underbrace{\underbrace{\underbrace{[Fe(OH)_3]_n \cdot mFeO^+}_{\text{胶核}} \cdot (m-x)Cl^-}_{\text{紧密层}}\}^{x+} \cdot \underbrace{xCl^-}_{\text{扩散层}}}_{\text{胶团}}$$

可滑动面

在外加电场作用下,分散相粒子产生定向移动,这种现象称为电泳。观察电泳现象,不仅可以确定胶粒所带电荷的符号,还可以计算胶粒的 ξ 电势。

ξ 电势是胶粒表面(即可滑动面)与溶液本体之间的电势差,只有在固液两相发生相对移动时,才呈现出 ξ 电势。测定 ξ 电势,对解决溶胶的稳定和聚沉问题有很大意义。

ξ 电势可由下式计算:

$$\xi = \eta u / \varepsilon E = \eta u / \varepsilon_r \varepsilon_0 E \qquad (3-13-1)$$

式中,E 为电场强度($E = \dfrac{V}{l}$),$V \cdot m^{-1}$;V 为外压电压,V;l 为两极间距离,m;ε 为分散介质的介电常数,$F \cdot m^{-1}$;ε_r 为介质的相对介电常数($\varepsilon = \varepsilon_r \varepsilon_0$),单位为 1;$\varepsilon_0 = 8.854\,2 \times 10^{-12} F \cdot m^{-1}$,为真空介质电常数;$\eta$ 为分散介质的黏度,$Pa \cdot s$;u 为电泳速度,$m \cdot s^{-1}$。

对于球形胶粒有 $\xi = \dfrac{6\pi\eta}{\varepsilon_r} \cdot \dfrac{u}{E}$,而对于棒状胶粒有 $\xi = \dfrac{4\pi\eta}{\varepsilon_r} \cdot \dfrac{u}{E}$,由于胶粒的特性不同以及电导辅助液与溶胶体系电导率的差异,因此须对有关公式进行修正,则

$$\xi = K \cdot \dfrac{\pi\eta}{\varepsilon_r} \cdot \dfrac{u}{E} \qquad (3-13-2)$$

式中,K 为与胶粒形状等有关的常数。对于球形胶粒,$K = 5.4 \times 10^{10}(V^2 \cdot s^2 \cdot kg^{-1} \cdot m^{-1})$;对于棒状胶粒 $K = 3.6 \times 10^{10}(V^2 \cdot s^2 \cdot kg^{-1} \cdot m^{-1})$。本实验可将 $Fe(OH)_3$ 溶胶视为棒状胶粒。

对于同一溶胶体系,若已知 η 和 ε_r(稀溶液可近似处理为水,也可以通过其他实验测定),实验测量 u 和 E,则可计算动电势 ξ。

2. 电解质对溶胶的聚沉作用

憎液溶胶中分散相粒子互相聚结,颗粒变大并发生沉降的现象称为聚沉。在溶胶中加入适量的电解质溶液,会引起溶胶发生聚沉作用,这是因为电解质的加入,使和分散相粒子所带电荷符号相反的离子(即反离子)进入了吸附层,而抵消了胶粒的电荷,故 ξ 电势降低,胶体的稳定性减小,致使溶胶产生聚沉。无机盐类是常用的聚沉剂,通常用聚沉值表示聚沉能力。在一定的条件下,能使某溶胶发生聚沉作用所需电解质的最低浓度称为该电解质对该溶胶的聚沉值。电解质的聚沉能力决定于与胶粒电荷符号相反的离子,且随着离子增加而增强。一般情况下,就聚沉能力(即聚沉值的倒数)而言,二价离子超过一价离子数十倍,而三价离子往往是一价离子的数百倍。

3. 乳状液的形成及其类型

一种液体以液滴的形式分散于另一种不相溶的液体中所形成的分散系统,称为乳状液。液滴的大小通常为 $1 \sim 50~\mu m$,因此可以在简单的显微镜下看到。

通常乳状液中一相是水,另一相是有机液体,习惯上称为油。乳状液可分为两种类型,一种为油分散于水中,称为水包油型,以 O/W 表示;另一种为水分散于油中,称为油包水型,以 W/O 表示。

乳状液的制备一般采用分散法,且必须加入第三种物质——乳化剂,在适当的条件下,使用不同的乳化剂可形成不同类型的乳状液。

鉴别乳状液类型的方法通常有三种,即电导法、稀释法及染色法。电导法是利用水和油的导电能力不同,水可以导电,油可以认为不导电,根据乳状液电导率的数量级即可判别其类型;稀释法是把两滴乳状液置于载破片上,分别滴加水和油,若能与水混溶则为 O/W 型,

如果与油混溶则为 W/O 型;染色法是利用有机染料溶于油不溶水的特性,取一滴乳状液置于载玻片上,加入少许只溶于油的染料如苏丹-Ⅲ,混匀后在显微镜下观察,若在无色连续相中分布着红色的小油滴则为 O/W 型,而当无色液滴分散在红色连续相内则为 W/O 型。当然也可以选用只溶于水的染料,情况正好相反。

在一定条件下可以把 O/W 型乳状液和 W/O 型乳状液相互转换,也就是在乳状液中进行转相。发生这种情况的原因之一可以是改变稳定剂的特性。如通过化学方法把碱金属皂转化为碱土金属皂,即可使 O/W 型乳状液转化为 W/O 型乳状液。

乳状液是热力学不稳定体系,其能长期稳定存在的一个重要原因是乳状液液珠带电。乳状液的液珠上所带电荷主要来自所使用的离子型表面活性剂在水相中电离产生的乳化剂离子在液珠表面的吸附所致。它以疏水的碳氢链伸入油相,而以离子头伸入水相的吸附态吸附于油水界面上,使形成的 O/W 型乳状液中的油珠带电。若乳化剂为阴离子型表面活性剂则油珠带正电荷,在带电的油珠周围还会有反离子呈扩散的状态分布,形成类似 Stern 模型的扩散双电层。由于双电层的相互排斥作用,使油珠之间不易接近,从而阻止了油珠的聚并。

乳状液的破坏是先发生絮凝,然后聚结,逐步被破坏,因而絮凝是液珠合并的前奏,这与液珠的长程力有关。胶体颗粒稳定的 DLVO 理论基础也适合于乳状液,即范德华力使得液珠颗粒相互吸引,当液珠接近到表面上的双电层发生相互重叠时,电排斥作用使液珠分开。如果排斥作用大于吸引作用,则液珠不易接触,因而不发生聚结,有利于乳状液的稳定。

本实验采用微电泳仪测定乳状液液珠的 ξ 电势。

【仪器与试剂】

仪器:电泳管 1 支,直流稳压电源(110 V),铜电极 2 个,电加热设备 1 套,JS94H 微电泳仪 1 台,显微镜 1 台,洗瓶 1 个,烧杯(250 mL)1 个,量筒(100 mL、10 mL)各 1 个,有塞量筒(50 mL)1 个,移液管(10 mL)2 支,试管 15 支,大试管 1 支,滴管 1 支。

药品:$w(FeCl_3) \approx 10\%$ 的 $FeCl_3$ 溶液,$0.001\ mol \cdot L^{-1}$ KCl 辅助溶液,$2.5\ mol \cdot L^{-1}$ 的 KCl 溶液,$0.1\ mol \cdot L^{-1}$ 的 K_2CrO_4 溶液,$0.1\ mol \cdot L^{-1}$ 的 $K_3[Fe(CN)_6]$ 溶液,煤油或菜籽油,1% 的油酸钠溶液,10% 的 $MgCl_2$ 溶液,苏丹-Ⅲ 染料。

【实验步骤】

1. 制备 $Fe(OH)_3$ 溶胶。在 250 mL 烧杯中放 95 mL 蒸馏水,加热至沸,慢慢地滴入 5 mL $w(FeCl_3) \approx 10\%$ 的 $FeCl_3$ 溶液,并不断搅拌,加完后继续沸腾几分钟,即得红棕色 $Fe(OH)_3$ 溶胶。为控制其浓度,将其放冷后,倾入量筒中,稀释到 100 mL。

2. 用界面移动法观测 $Fe(OH)_3$ 溶胶的电泳,计算 ξ 电势。

(1) 电泳测量仪的准备。

①电泳测量仪:电泳仪应事先用铬酸洗液洗涤清洁,以除去管壁上可能存在的杂质,然后用水洗净并烘干;活塞上涂一薄层凡士林,塞好活塞待用。

②电导辅助液的配制:将制备并经净化的溶胶置入电导池中,测量其电导率;另取 300 mL 蒸馏水,插入洁净的电导电极,一边搅拌一边慢慢滴加 $0.001\ mol \cdot L^{-1}$ 的 KCl 溶液,测量其电导率值,直至与溶胶的电导率值相同为止。

③电泳仪测量液体的安装:把制得纯化的 Fe(OH)₃ 待测溶胶溶液通过长颈小漏斗缓慢地注入电泳仪的 U 形管底部,至适当高度位置。将电泳仪垂直安置稳固,静置。用滴管将电导辅助溶液,分别沿 U 形管左右管壁徐徐加入,先少后多,先慢后快,直至约 10 cm 高度,调整两边的总量相等并且等高。须注意要始终保持两种溶液的液接面之间的界面清晰。

(2) Fe(OH)₃ 溶胶的 ξ 电势测定。

轻轻将铂电极插入 KCl 液层中并固定。切勿扰动页面,铂电极应保持垂直,并使两极进入液面下的深度相等,记下胶体液面的高度位置。按图 3-13-1 所示,连接好电路。将两电极连接到输出为 30~50 V 的直流电源上,或者直接连接于专用电泳仪直流稳压电源上,调整好所需电压。

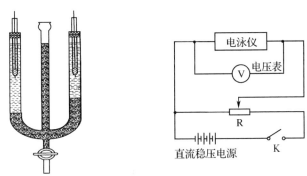

图 3-13-1 电泳仪及测 ξ 电势电路图

接通电源,按动秒表开始计时,约 30~45 min 时,记下胶体溶液上升变化的距离 d 和外加电压的读数;沿 U 形管中线量出电极间的距离,此数值须测量多次,并取其平均值为 L,也可将电压反向连接到电极上,重复测量电泳的距离和电压。实验结束后应回收胶体溶液,并在 U 形管中放满水浸泡铂电极。

3. 电解质对溶胶的聚沉作用。

用 10 mL 移液管在三个干净的 50 mL 锥形瓶中各注入 10 mL 前面用水解法制备的 Fe(OH)₃ 溶胶,然后在每个瓶中用滴定管一滴滴地慢慢加入 0.5 mol·L⁻¹ KCl, 0.01 mol·L⁻¹ K_2SO_4, 0.001 mol·L⁻¹ $K_3Fe(CN)_6$ 溶液,不断摇动。注意,在开始有明显聚沉物出现时,即停止加入电解质,记下每次所用溶液的毫升数。

4. 制备乳状液并鉴别其类型。

取 20 mL 水与 10 mL 煤油于有塞量筒中,剧烈振荡,观察现象。

取 20 mL 1% 油酸钠溶液于有塞量筒中,再加入 10 mL 煤油(或菜籽油),盖以瓶塞并剧烈震荡,观察乳状液的形成,并用稀释法和染色法观察乳状液的类型。

取 5 mL 乳状液于试管中,然后加入数滴氯化镁溶液,用玻璃棒充分搅拌,再用染色法鉴别其类型。

5. 用 JS94H 微电泳仪测定乳状液的 ξ 电势。

打开微电泳仪,预热 10 min,并将仪器与电脑连接。将上述制得的乳状液稀释至可透光程度,取少量于电泳杯中。用十字标调好焦距后,换上铂电极,测定乳状液的 ξ 电势。JS94H 微电泳仪的使用见附录。

【实验记录及数据处理】

1. 计算聚沉值大小,说明溶胶带什么电?与理论值比较,说明什么问题?
2. 由 U 形管内溶胶在时间 t 内界面上升移动的距离 d 值,计算电泳的速度 $u = d/t$。
3. 由测得的电压 U 和两极间的距离 L,计算出电位梯度 $E = U/L$。
4. 将 u 和 E 代入式(3-13-2)计算出 $Fe(OH)_3$ 溶胶胶粒的 ξ 电势。

【讨论与说明】

1. 测电泳时两溶液之间的界面必须清晰,否则需重新装样。
2. 反复测量时,两液之间的界面清晰度将被破坏,只要明显可辨明,就可继续(或使用不同的电压值)进行实验。
3. 电源电压的数值在实验的进程中也应读数并记录。

【思考题】

1. 如果事先电泳仪没有洗净,管壁上残留有微量的电解质,对电泳测量的结果有什么影响?
2. 电泳速度的快慢与哪些因素有关?

【相关阅读】

ξ 电势与溶胶体系的稳定性

溶胶是高度分散的多相系统,具有较高的表面能,是热力学不稳定系统,因此溶胶粒子有自动聚集变大的趋势。事实上很多溶胶可以在相当长的时间内稳定存在,主要原因有以下几方面:

1. 动力学稳定性。溶胶体系是高度分散的,分散相粒子较小,布朗运动激烈,在重力场中不易沉降,使溶胶具有动力学稳定性。
2. 溶剂化稳定性。胶体粒子是水化的,相当于胶粒外围有一层水膜,水化层中的水分子定向排列,两胶粒若靠近,水分子的弹性产生机械阻力,使两胶粒不易靠近;另一方面,胶粒若絮凝成大粒子,须先把水化层去掉,这需要额外的能量。这些都有利于溶胶稳定。
3. 带电稳定性。胶粒带电是溶胶稳定性的最重要原因,目前解释溶胶稳定性较完善的理论是 DLVO 理论。

20 世纪 40 年代,苏联学者 Deijaguin 和 Landau 与荷兰学者 Verwey 和 Overbeek 分别提出了带电胶体粒子稳定的理论,该理论以各位科学家姓名的首字母组合命名,称为 DLVO 理论,其要点如下:

(1) 分散在介质中的胶粒之间既存在引力又存在斥力。胶粒间的吸引力在本质上和分子间的范德华力相似,只是比一般分子间范德华力大千百倍之多,称为远程范德华力。胶粒间的引力势能与粒子间距离的一次方或二次方成反比。当两个胶粒间距离较大时,双电层未重叠,主要表现吸引力;随胶粒间距离的减小,双电层部分重叠,在重叠区内反离子的浓度增加,导致两个胶团扩散层的对称性同时遭到破坏,这样既破坏了扩散层中反离子的平衡分布,也破坏了双电层的静电平衡,前一平衡的破坏使重叠区内过剩的反离子向未重叠区扩散,导致渗透性斥力的产生、平衡的破坏,进而导致两胶团间产生静电斥力。随重叠区的增大,斥力势能增加。

(2) 溶胶系统的稳定和聚沉取决于引力势能和斥力势能的相对大小。如果斥力势能大于引力势能,则凝胶处于相对稳定的状态;如果引力势能大于斥力势能,凝胶会发生聚沉。

(3) 斥力势能、引力势能和总势能随胶粒间距离变化而变化,其关系如图3-13-2所示。当胶粒间距离较大时,吸引力起主要作用,势能是负值;胶粒靠近时,双电层逐渐重叠,排斥力起主要作用,势能为正值,且使总势能曲线出现最大值。此最大值代表溶胶发生聚沉时必须克服的势垒,当胶粒的动能不足以越过此势垒时,溶胶能稳定存在,若胶粒动能可超越此势垒,将导致

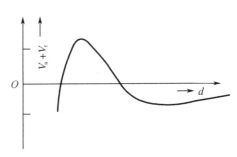

图 3-13-2　粒子间作用能与其距离的关系曲线

溶胶聚沉;当胶粒距离缩短到一定程度后,吸引力又占主导地位,位能为负值,胶粒距离在此范围可形成结构紧密而又稳定的聚沉物。

溶胶的带电稳定性是溶胶稳定的决定因素,而溶胶的 ξ 电势是其胶粒表面电荷密度大小的反映,是衡量溶胶稳定性的重要标志。

很多物质与极性介质(如水)接触后,其界面上就带有电荷。界面可以通过一种或几种机理获得电荷,最为常见的机理包括:①表面离子的优先(或不等量)溶解;②表面的直接电离;③表面离子的取代;④特殊离子的吸附;⑤从特殊晶体结构中产生电荷等。

以溶胶粒子选择性地吸附某种离子而带电为例,吸附正离子胶粒带正电,吸附负离子带负电,但整个溶液是电中性的,故还应有等量的符号相反的离子(反离子)存在。固粒表面吸附的离子和溶液中的反离子构成双电层。反离子在溶液中受到两个方向相反的作用:
①固体粒子表面被吸附的离子的引力,力图将它们(反离子)拉向界面;②离子本身的热运动使反离子离开界面而扩散到溶液中去,其结果使反离子在固粒表面外呈平衡分布,靠近界面处反离子浓度大些。随着与界面距离的增大,反离子由多到少,形成扩散分布,如图3-13-3所示。MN 代表粒子的平表面,设它吸附正离子,则电量相等的负离子扩散分布,就好像大气层中气体分子按高度分布的状态。直到界面正电荷电力所不及处,过剩离子浓度等于零。带电表面及这些反离子构成扩散双电

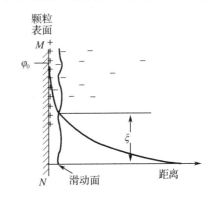

图 3-13-3　扩散双电层

层。双电层的厚度随溶液中离子浓度和电荷数而不同。

粒子带电是溶胶相对稳定的重要因素。

实验证明,溶胶粒子在电场作用下与溶液发生相对移动时,分界面不是在固液界面 MN 处,而是有一层液体牢固地附在固体表面,即滑动面以内部分,并随表面运动。按电泳算出

的就是此滑动面上的电势,称电动电势或ξ电势。

ξ电势的大小与吸附层中的离子以及扩散层厚度有关,受外加电解质的影响很大。随着外加电解质浓度的增加,将使更多的反离子进入吸附层中,同时压缩了扩散层的厚度,当扩散层被压缩到与吸附层相重叠时,ξ电势降低到零,这种状态为等电状态,这时胶粒不带电荷,易发生沉聚。如果外加电解质中反离子发生强烈的专性吸附,甚至可以使ξ电势的符号改变。

实验14　黏度法测定高聚物的摩尔质量

【实验目的】

1. 掌握用乌氏(Ubbelohde)黏度计测定高聚物溶液黏度的原理和方法;
2. 明确增比黏度、比浓黏度、特性黏度的概念;
3. 掌握测定线型高聚物聚乙二醇的相对分子质量的平均值。

【预习要求】

1. 了解乌氏黏度计的构造与使用方法;
2. 理解高聚物相对分子量、黏度、相对黏度、增比黏度、比浓黏度、特性黏度等概念。

【实验原理】

黏度是指液体流动时所表现出来的阻力,这种力反抗液体中邻接部分的相对移动,因此可看作是一种内摩擦力。若在两平行板间盛以某种液体,一块静止,另一块以速度 v 向 x 方向做匀速运动。如果将液体沿 y 方向分成很多薄层,则各液层沿 x 方向的流速随 y 值的不同而变化,如图 3-14-1 所示,流体的这种形变称为切变。流体流动时有速度梯度存在,运动较慢的液层阻滞较快的液层的运动,因此产生流动阻力。为了维持稳定的流动,保持速度梯度不变,则要对上面的平板施加恒定的力(切力)。若平板的面积是 A,则切力为

$$f = \eta A \frac{\mathrm{d}v}{\mathrm{d}y}$$

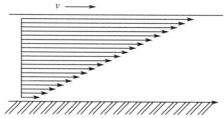

图 3-14-1　两平面间的黏性流动

式中,η 称为该液体的黏度系数(简称黏度),单位为 Pa·s,上式称为牛顿黏度公式。符合

牛顿黏度公式的液体称为牛顿流体。

单体分子经加聚或缩聚过程便可合成高聚物。高聚物并非每个分子的大小都相同,即聚合度不一定相同,所以高聚物的摩尔质量是一个统计平均值。对于聚合物研究来说,高聚物平均摩尔质量是必须测定的重要数据之一,平均摩尔质量根据平均的方法不同分为数均摩尔质量、质均摩尔质量、Z均摩尔质量、黏均摩尔质量。每种平均摩尔质量可通过各种相应的物理或化学方法进行测定。

高聚物稀溶液的黏度,主要反映了液体在流动时存在着内摩擦。其中因溶剂分子之间的内摩擦表现出来的黏度叫纯溶剂黏度,记作 η_0;此外还有高聚物分子相互之间的内摩擦以及高分子与溶剂分子之间的内摩擦。三者之总和表现为溶液的黏度 η。在同一温度下,一般来说,$\eta > \eta_0$。相对于溶剂,其溶液黏度增加的分数称为增比黏度,记作 η_{sp},即

$$\eta_{sp} = \frac{\eta - \eta_0}{\eta_0}$$

而溶液黏度与纯溶剂黏度的比值称为相对黏度,记作 η_r,即

$$\eta_r = \frac{\eta}{\eta_0}$$

η_r 也是整个溶液的黏度行为,η_{sp} 则意味着已扣除了溶剂分子之间的内摩擦效应。两者关系为

$$\eta_{sp} = \frac{\eta}{\eta_0} - 1 = \eta_r - 1$$

对于高分子溶液,增比黏度 η_{sp} 往往随溶液的浓度 C 的增加而增加。为了便于比较,将单位尝试下所显示出的增比黏度,即 η_{sp}/C 称为比浓黏度;而 $\ln\eta_r/C$ 称为比浓对数黏度。η_r 和 η_{sp} 都是无因次的量。

为了进一步消除高聚物分子间内摩擦的作用,将溶液无限稀释,当浓度 C 趋近于零时,高聚物分子之间相隔较远,它们之间的作用可以忽略,比浓黏度和比浓对数黏度趋近于一个极限值,即

$$\lim_{c \to 0} \frac{\eta_{sp}}{C} = \lim_{c \to 0} \frac{\eta_r}{C} = [\eta]$$

$[\eta]$ 主要反映了高聚物分子与溶剂分子之间的内摩擦作用,称之为高聚物溶液的特性黏度。由于 η_{sp} 和 η_r 均是无量纲量,所以 $[\eta]$ 的单位是浓度 C 的倒数。在文献和实验教材中 C 及 $[\eta]$ 所用单位不尽相同,本实验 C 的单位为 $kg \cdot m^{-1}$,$[\eta]$ 的单位为 $kg^{-1} \cdot m^3$,其数值可通过实验求得。在足够稀的溶液中有

$$\frac{\eta_{sp}}{C} = [\eta] + \kappa[\eta]^2 C$$

$$\frac{\ln \eta_{sp}}{C} = [\eta] - \beta[\eta]^2 C$$

上面两式中 κ 和 β 分别称为 Huggins 常数和 Kramerr 常数。这样,以 η_{sp}/C 对 C 作图、以 $\ln \eta_r/C$ 对 C 作图可得两条直线,对于同一高聚物,这两条直线在纵坐标轴上相交于一点,如图 3-14-2 所示,可求出 $[\eta]$ 数值。由溶液的特性黏度 $[\eta]$ 还无法获得高聚物的摩尔质量数据,目前高聚物溶液的特性黏度 $[\eta]$ 与高聚物平均相对摩尔质量之间的关系通常由半经验的麦克(H. Mark)经验方程式来求得

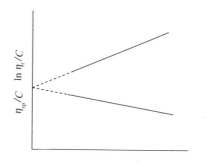

图 3-14-2 外推法求 $[\eta]$ 图

$$[\eta] = K \overline{M}_\eta^\alpha$$

式中,K 为比例系数,α 是与分子形状有关的经验常数。它们都与温度、聚合物和溶剂性质有关,在一定的相对分子质量范围内与相对分子质量无关。

K 和 α 的数值只能通过其他方法确定,如渗透压法、光散射法等。黏度法只能测定 $[\eta]$ 并求算出 $\overline{M}$。对于聚乙二醇的水溶液,不同温度下的 K、α 值如表 3-14-1 所示。

综上所述,溶液黏度的名称、符号及定义可归纳为表 3-14-2。

表 3-14-1 聚乙二醇不同温度时的 K、α 值(溶剂为水)

$t/℃$	$K/(10^{-6} \cdot m^3 \cdot kg^{-1})$	α	$\overline{M}_\eta$
25	156	0.50	0.091~0.1
30	12.5	0.78	2~500
35	6.4	0.82	3~700
40	16.6	0.82	0.04~0.4
45	6.9	0.81	3~700

表 3-14-2 溶液黏度的命名

名称	符号和定义
黏度(系数)	η
相对黏度	$\eta_r = \eta/\eta_0$(η_0 为溶剂的黏度)
增比黏度	$\eta_{sp} = \eta_r - 1 = (\eta - \eta_0)/\eta_r$
比浓黏度	η_{sp}/C
比浓对数黏度	$\ln \eta_r/C$
特性黏度	$[\eta] = (\eta_{sp}/C)_{c \to 0} = (\ln \eta_r/C)_{c \to 0}$

测定液体黏度的方法在前文中已经介绍,本实验采用毛细管法测定黏度,通过测定一定体积的液体流经一定长度和半径的毛细管所需时间而获得,本实验使用乌式黏度计。当液体在毛细管黏度计内因重力作用而流出时遵守泊肃叶(Poiseuille)定律,即

$$\frac{\eta}{\rho} = \frac{\pi h g r^4 t}{8Vl} - m\frac{V}{8\pi l t}$$

式中,ρ 为液体的密度;l 是毛细管长度;r 是毛细管半径;t 是流出时间;h 是流经毛细管液体的平均液柱。$l \ll 1$ 时,可取 $m = 1$。

对某一支指定的黏度计而言,令 $\alpha = \frac{\pi h g r^4}{8Vl}$,$\beta = \frac{mV}{8\pi l t}$,则上式可改写为

$$\frac{\eta}{\rho} = \alpha t - \frac{\beta}{t}$$

式中当 $\beta < 1$,$t > 100$ s 时,等式右边第二项可以忽略。设溶液的密度 ρ 与溶剂密度 ρ_0 近似相等。这样,通过分别测定溶液和溶剂的流出时间 t 和 t_0,就可求算 η_r,即

$$\eta_r = \frac{\eta}{\eta_0} = \frac{t}{t_0}$$

进而可分别计算得到 η_{sp}、η_{sp}/C 和 $\ln\frac{\eta_r}{C}$ 值。配置一系列不同浓度的溶液分别进行测定,以 η_{sp}/C 和 $\ln\frac{\eta_r}{C}$ 为同一纵坐标,C 为横坐标作图,得两条直线,分别外推到 $C = 0$ 处(如图 3-14-2 所示),其截距即为 $[\eta]$,代入 $[\eta] = K\overline{M}_\eta^\alpha$,即可得到。

【仪器与试剂】

恒温水浴 1 套,乌氏黏度计 1 支(如图 3-14-3 所示),具塞锥形瓶(50 mL)2 只,洗耳球 1 只,移液管(5 mL)1 支,移液管(10 mL)2 支,细乳胶管 2 根,恒温槽夹 1 个,容量瓶(25 mL)1 只,秒表 1 只,弹簧夹及聚乙二醇(分析纯)。

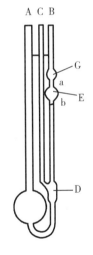

图 3-14-3　乌氏黏度计

【实验步骤】

1. 根据室温,将恒温水浴调至 25 ± 0.1 ℃,30 ± 0.1 ℃ 或 35 ± 0.1 ℃。

2. 溶液配制。称取聚乙二醇 1 g(称准至 0.001 g),在 25 mL 容量瓶中配成水溶液。配溶液时,要先加入溶剂至容量瓶的 2/3 处,待其全部溶解后恒温 10 min,再用同样温度的蒸馏水稀至刻度。

3. 洗涤黏度计。先用热洗液(经砂芯漏斗过滤)浸泡,再用自来水、蒸馏水冲洗,经常使用的黏度计则用蒸馏水浸泡,去除留在黏度计中的高聚物。黏度计的毛细管要反复用水冲洗。

4. 测定溶剂流出时间 t_0。将黏度计垂直夹在恒温水浴内,将 10 mL 纯溶剂自 A 管注入黏度计内,恒温数分钟,夹紧 C 管上连接的乳胶管,同时在连接 B 的乳胶管上慢慢抽气,待液体升至 G 球的一半左右即停止抽气,打开 C 管乳胶管上的夹子,使毛细管内液体同 D 球中

的液体分开,用秒表测定液面在 a、b 两线间移动所需时间。重复测定三次,每次相差不超过 $0.2 \sim 0.3$ s,取平均值。

5. 测定溶液流出时间 t。取出黏度计,倒出溶剂,吹干。用移液管吸取 10 mL 已恒温的高聚物溶液,同上法测定流经时间 t。

6. 再用移液管加入 50 mL 已恒温的溶剂(改变试样浓度),用洗耳球从 C 管鼓气搅拌并将溶液慢慢地抽上流下数次使之混合均匀,再如上法测定流经时间 t。同样,再依次加入 5 mL、10 mL、15 mL 溶剂,改变试样浓度,逐一测定不同浓度溶液的流经时间。

7. 实验结束后,将溶液倒入回收瓶内,用溶剂仔细冲洗黏度计三次,最后用溶剂浸泡,备用。

【实验记录及数据处理】

1. 根据实验数据与不同浓度的溶液测定的相应的流出时间,计算 η_r、η_{sp}、η_{sp}/C 和 $\ln \eta_r/C$。
2. 以 η_{sp}/C 及 $\ln \eta_r/C$ 分别对 C 作图,得到两条直线,并外推至 $C=0$ 处,求得 $[\eta]$。
3. 利用实验温度下取得的常数 K、α 值,按式 $[\eta]=K\overline{M}_\eta^\alpha$ 计算出聚乙二醇的黏均摩尔质量 $\overline{M}_\eta$。

【讨论与说明】

1. 黏度计必须洁净,如毛细管壁上挂有水珠,需用洗液浸泡。
2. 高聚物在溶剂中溶解缓慢,配制溶液时必须保证其完全溶解,否则会影响溶液起始浓度,而导致结果偏低。
3. 本实验中溶液的稀释是直接在黏度计中进行的,所用溶剂必须先在与溶液同处的恒温水浴中恒温,然后用移液管准确量取并充分混合均匀方可测定。
4. 测定时黏度计要垂直放置,否则会影响结果的准确性。

【思考题】

1. 乌氏黏度计中的 C 管的作用是什么?能否去掉 C 管改为双管黏度计,为什么?
2. 黏度计的毛细管太粗或太细有什么缺点?
3. 黏度法测定高聚物相对分子质量的优点是什么?黏度法测定高聚物相对分子质量有哪些局限性?

【相关阅读】

聚合物相对分子质量及相对分子质量分布

聚合物的相对分子质量较一般有机化合物的相对分子质量要大得多,且聚合物相对分子质量具有不均一性(多分散),即单个聚合物分子间的相对分子质量不相同。宏观的聚合物相对分子质量只是所有单个聚合物分子不同相对分子质量的一个平均值,单个聚合物分子间相对分子质量的不均一(分散)程度由相对分子质量分布来表达和描述。聚合物的许多独特性质都与其相对分子质量及其相对分子质量分布有关,如分子链的柔顺性、聚合物的熔点、玻璃化温度、黏度以及力学性能等。

1. 相对分子质量及相对分子质量分布的定义

根据统计平均方法的不同,统计平均相对分子质量常用的方法有以下四种。

(1) 数均相对分子质量,按聚合物分子数统计平均,即

$$\overline{M_n} = \frac{\sum N_i M_i}{\sum N_i} = \sum n_i M_i \qquad (3-14-1)$$

式中,N_i 为相对分子质量为 M_i 的物质的量,n_i 为相对分子质量为 M_i 的摩尔分数。

(2) 质均相对分子质量,按聚合物质量统计平均,即

$$\overline{M_w} = \frac{\sum W_i M_i}{\sum W_i} = \sum w_i M_i \qquad (3-14-2)$$

式中,W_i 为相对分子质量为 M_i 的物质的量,w_i 为相对分子质量为 M_i 的摩尔分数。

(3) Z 均相对分子质量,按 Z 量统计平均,即

$$\overline{M_Z} = \frac{\sum (W_i M_i) M_i}{\sum W_i M_i} = \frac{\sum W_i M_i^2}{\sum W_i M_i} \qquad (3-14-3)$$

式中,W_i 为相对分子质量为 M_i 的质量。

(4) 黏均相对分子质量,根据 $[\eta] = kM^\alpha$,得

$$\overline{M_\eta} = \left(\frac{\sum N_i M_i^{\alpha+1}}{\sum W_i M_i} \right)^{1/\alpha} \qquad (3-14-4)$$

式中,α 为常数,取值为 $0.5 \sim 0.8$,取决于温度和具体的聚合物与溶剂的组合,取聚合物链段和溶剂分子间热力学的相互作用。

在四种平均相对分子质量表达方法中,数均相对分子质量和质均相对分子质量最常使用,Z 均相对分子质量与黏均相对分子质量由于物理意义不太明确,应用较少。

由于聚合物相对分子质量的多分散特性,仅用相对分子质量的平均值尚不能完全反映聚合物相对分子质量的真实情况,尚需相对分子质量分布描述聚合物相对分子质量的分散情况。

聚合物的相对分子质量分布有多种表达方法,最简便也最常用的表达方法是采用质均相对分子质量与数均相对分子质量的比值,即

$$D = \overline{M_W}/\overline{M_n}$$

2. 聚合物相对分子质量及相对分子质量分布的测定方法

测定聚合物相对分子质量及相对分子质量分布的方法有许多种,常用的方法有端基滴定法、冰点降低法、蒸汽压渗透法、膜渗透压法、光散射法、黏度法和色谱法,如表 3-14-3 所示。

端基滴定法是化学方法,通过聚合物端基与滴定试剂的化学反应完成测试;冰点降低法、蒸汽压渗透法、膜渗透压法根据聚合物稀溶液的热力学性质确定聚合物相对分子质量;

光散射法的基本原理是利用光线在聚合物溶液中的散射实验来进行测定的;黏度法测定相对分子质量所依据的是聚合物和溶液小分子化合物之间的黏度差别;色谱法可得到聚合物的相对分子质量分布。

表 3-14-3 相对分子质量测定方法

测试方法	相对分子质量范围	相对分子质量
端基滴定法	3×10^4 以下	数均
冰点降低法	3×10^3 以下	数均
蒸汽压渗透法	3×10^4 以下	数均
膜渗透压法	$3 \times 10^4 \sim 1.5 \times 10^6$	数均
光散射法	$1 \times 10^4 \sim 1 \times 10^7$	质均
黏度法	$1 \times 10^4 \sim 1 \times 10^7$	黏均
色谱法	$1 \times 10^2 \sim 1 \times 10^7$	各种

由上述各种方法,根据不同的测定原理,可得到数均、质均、黏均相对分子质量。

测试方法又有绝对法和相对法之分。如果测定的物理量与相对分子质量有直接的理论关系,通过测定这个物理量就得到了绝对分子质量,则为绝对法;如果测定的物理量与相对分子质量的关系需要进一步校正,则为相对法,得到的就是相对分子质量。

聚合物的相对分子质量分布的测定,目前几乎完全采用色谱法,可直接给出多种聚合物相对分子质量分布的表达形式。

实验 15　溶液吸附法测定固体比表面积

【实验目的】

1. 学会用次甲基蓝水溶液吸附法测定活性炭的比表面积;
2. 了解朗缪尔(Langmuir)单分子层吸附理论及溶液法测定比表面积的基本原理。

【预习要求】

1. 了解比表面的概念;
2. 了解朗缪尔单分子层吸附理论;
3. 了解溶液法测定比表面积的基本原理。

【实验原理】

测定固体比表面积的方法很多,有 BET 低温吸附法、气相色谱法、电子显微镜法等。这些方法需要较复杂的仪器装置或较长的实验时间。相比之下,溶液吸附法测定固体比表面积具有仪器装置简单、操作方便等优点。

水溶性染料的吸附可用于测定固体比表面积。在所有的染料中,次甲基蓝是易于被固体吸附的水溶性染料。研究表明,在一定浓度范围内,大多数固体对次甲基蓝的吸附是单分子层吸附,符合郎缪尔吸附理论。

郎缪尔吸附理论的基本假设是:固体表面是均匀的,吸附是单分子层吸附,吸附剂一旦被吸附质覆盖就不能被再吸附;在吸附平衡时候,吸附和脱附建立动态平衡;吸附平衡前,吸附速率与空白表面成正比,解吸速率与覆盖度成正比。

设固体表面的吸附位总数为 N,覆盖度为 θ,溶液中吸附质的浓度为 C,根据上述假定,有

吸附速率: $\quad r_{吸} = k_1 N(1-\theta)C$ (k_1 为吸附速率常数)

脱附速率: $\quad r_{脱} = k_{-1} N\theta$ (k_{-1} 为脱附速率常数)

当达到吸附平衡时,吸附速率与脱附速率相等,即

$$k_1 N(1-\theta)C = k_{-1} N\theta$$

由此可得

$$\theta = \frac{K_{吸} C}{1 + K_{吸} C} \qquad (3-15-1)$$

式中,$K_{吸} = k_1/k_{-1}$ 称为吸附平衡常数,其值决定于吸附剂和吸附质的性质及温度,$K_{吸}$ 值越大,固体对吸附质吸附能力越强。若以 Γ 表示浓度 C 时的平衡吸附量,以 Γ_∞ 表示全部吸附位被占据时单分子层吸附量,即饱和吸附量,将 $\theta = \Gamma/\Gamma_\infty$ 带入式(3-15-1)得

$$\Gamma = \Gamma_\infty \frac{K_{吸} C}{1 + K_{吸} C} \qquad (3-15-2)$$

整理式(3-15-2)得到如下形式

$$\frac{C}{\Gamma} = \frac{1}{\Gamma_\infty K_{吸}} + \frac{1}{\Gamma_\infty} C \qquad (3-15-3)$$

而计算吸附量可由平衡浓度 C 及初始浓度 C_0 数据,按式(3-15-4)计算吸附量 Γ。

$$\Gamma = \frac{(C_0 - C)V}{m} \qquad (3-15-4)$$

式中,$V(L)$ 为吸附溶液的总体积,$m(g)$ 为加入溶液的吸附剂质量。通过计算的 Γ 值,以 $C/\Gamma - C$ 作图,从直线斜率可求得 Γ_∞,再结合截距便可得到 $K_{吸}$。Γ_∞ 指每克吸附剂对吸附质的饱和吸附量(用物质的量表示),若每个吸附质分子在吸附剂上所占据的面积为 σ_A,则吸附剂的比表面积可以按照下式计算

$$S = \Gamma_\infty L \sigma_A \qquad (3-15-5)$$

式中,S 为吸附剂比表面积,L 为阿伏伽德罗常数。

次甲基蓝的结构为

$$\left[\begin{matrix} H_3C \\ H_3C \end{matrix}\right. \!\! N \!\!-\!\! \underset{S}{\diagdown}\!\!-\!\! N \!\! \left.\begin{matrix} CH_3 \\ CH_3 \end{matrix}\right]^+ Cl^-$$

阳离子大小为 $17.0 \times 7.6 \times 3.25 \times 10^{-30}$ m³。

次甲基蓝的吸附有三种取向：平面吸附投影面积为 135×10^{-20} m²，侧面吸附投影面积为 75×10^{-20} m²，端基吸附投影面积为 39×10^{-20} m²。对于非石墨型的活性炭，次甲基蓝是以端基吸附取向吸附在活性炭表面，因此 $\sigma_A = 39 \times 10^{-20}$ m²。由此可见，对溶液法吸附而言，非球形的吸附质在各种吸附剂表面吸附时的取向并非完全一致，每个吸附质分子的投影面积可以相差甚远，特别是三种吸附取向都存在时，所计算出来的结果与真实的比表面积会有较大的误差。溶液吸附法的测量误差通常为 10% 甚至更高。更加准确的比表面积的测量需用其他方法。

根据光吸收定律，当入射光为一定波长的单色光时，某溶液的吸光度与溶液中有色物质的浓度及溶液层的厚度成正比，即

$$A = -\lg(I/I_0) = \varepsilon bC \qquad (3-15-6)$$

式中，A 为吸光度，I_0 为入射光强度，I 为透过光强度，ε 为吸光系数，b 为光径长度或液层厚度，C 为溶液浓度。

次甲基蓝溶液在可见区有两个吸收峰：445 nm 和 665 nm。但在 445 nm 处活性炭吸附对吸收峰有很大的干扰，故本试验选用的工作波长为 665 nm，并用分光光度计进行测量。

【仪器与试剂】

722 型分光光度计及其附件 1 套，容量瓶（500 mL）6 只，容量瓶（100 mL）5 只，容量瓶（50 mL）5 只，HY 振荡器 1 台，2 号砂芯漏斗 5 只，带塞锥形瓶 5 只，滴管 2 支，次甲基蓝溶液（0.2% 左右原始溶液），次甲基蓝标准液（0.3126×10^{-3} mol·L⁻¹），颗粒状非石墨型活性炭，0.0001 g 电子天平 1 台，马弗炉 1 台。

【实验步骤】

1. 样品活化

将颗粒活性炭置于瓷坩埚中放入 500 ℃ 马弗炉活化 1 h，然后置于干燥器中备用。（此步骤实验前已经由实验室做好）

2. 溶液吸附

取五只干燥的带塞锥形瓶，编号为 1～5 号，分别准确称取活化过的活性炭约 0.1 g 置于瓶中。用移液管分别量取浓度为 0.2% 左右次甲基蓝原始溶液 30 mL、20 mL、15 mL、10 mL、5 mL 置于五只 50 mL 容量瓶中，加入蒸馏水稀释至刻度。把五只容量瓶中稀释后的次甲基蓝溶液分别倒入五只锥形瓶中，塞好塞子，放在振荡器上震荡 3 h。样品振荡达到平衡后，将锥形瓶取下，用砂芯漏斗过滤，得到吸附平衡后滤液。分别量取滤液 5 mL 于 500 mL 容量瓶中，用蒸馏水定容摇匀待用。此为平衡稀释液。

3. 原始溶液处理

为了准确测量 0.2% 次甲基蓝原始溶液的浓度，量取 2.5 mL 溶液放入 500 mL 容量瓶中，并用蒸馏水稀释至刻度，待用。此为原始溶液稀释液。

4. 次甲基蓝标准溶液的配制

分别量取 0.5 mL、1 mL、2 mL、4 mL、8 mL、12 mL 浓度为 0.3126×10^{-3} mol·L⁻¹ 的标准溶液于 100 mL 容量瓶中，加蒸馏水稀释至刻度，即得浓度依次为 0.005×10^{-3} mol·L⁻¹、

0.01×10^{-3} mol·L^{-1}、0.02×10^{-3} mol·L^{-1}、0.04×10^{-3} mol·L^{-1}、0.08×10^{-3} mol·L^{-1}、$0.12 \times (0.3126 \times 10^{-3}$ mol·L$^{-1})$ 溶液,分别编号为 A1、A2、A3、A4、A5、A6,配制好溶液后待用。

5. 选择工作波长

次甲基蓝溶液的工作波长为 665 nm。由于各分光光度计波长刻度略有误差,取步骤 4 中的 A4 号标准溶液,在 600~700 nm 范围内测量吸光度,以吸光度最大的波长为工作波长。

6. 测量吸光度

选择透光率 $T\%$ 的比色皿用作参比。因为次甲基具有吸附性,所以应按照从稀到浓的顺序测定。

在工作波长下,依次分别测定步骤 4 中标准溶液 A1~A6 及步骤 2 和步骤 3 中所配次甲基蓝溶液的吸光度。

【实验记录及数据处理】

1. 作次甲基蓝溶液吸光度对浓度的工作曲线。
2. 求次甲基蓝原始溶液浓度和各个平衡溶液浓度。

据稀释后原始溶液的吸光度,从工作曲线上查得对应的浓度,乘以相应的稀释倍数,即为原始溶液的浓度 C_0。

将试验测定的各个稀释后的平衡溶液吸光度,从工作曲线上查得对应的浓度,乘上稀释倍数,即为平衡溶液的浓度 C。

3. 计算吸附量 由平衡浓度 C 及初始浓度 C_0 数据,按式(3-15-4)计算吸附量 Γ。利用求得的 Γ,作 $\Gamma - C$ 吸附等温线。

4. 由 Γ 和 C 数据计算 C/Γ 值,然后作 $C/\Gamma - C$ 图,由图求得饱和吸附量 Γ_∞。将 Γ_∞ 值用虚线作一水平线在 $\Gamma - C$ 图上。这一虚线即是吸附量 Γ 的渐近线。

5. 计算试样的比表面积,求平均值。

【讨论与说明】

1. 活性炭颗粒要均匀并干燥。
2. 标准溶液的浓度要准确配制。
3. 振荡时间要充足,一般控制在 3 h 以上使其达到饱和。
4. 按朗缪尔吸附等温线的要求,溶液吸附必须在等温的条件下进行,将吸附瓶置于恒温水浴中进行振荡,使之达到平衡。但通常情况下室温变化并不显著,因而本实验仅在室温条件下将吸附瓶置于振荡器上振荡,如果实验期间室温变化过大,必然影响结果。

【思考题】

1. 根据本实验测得的吸附量能否画出对应的等温吸附曲线?
2. 溶液产生吸附时,如何判断其达到平衡?
3. 测定次甲基蓝原溶液和平衡溶液时,为什么要将溶液稀释?

【相关阅读】

比 表 面

比表面又称比表面积,系指 1 g 固体物质所具有的表面积。比表面包括内表面积和外表面积之和,单位为 $m^2 \cdot g^{-1}$,是表征固体性能的最重要的物化参数之一。因而,比表面积测定是生产、科研和教学工作中不可或缺的实验。

比表面积测试方法有两种分类标准:一是根据测定样品吸附气体量多少,可分为连续流动法、容量法及重量法(重量法现在基本上很少采用);另一种是根据计算比表面积理论方法不同可分为直接对比法比表面积分析测定、Langmuir 法比表面积分析测定和 BET 法比表面积分析测定等。同时这两种分类标准又有一定的联系,直接对比法只能采用连续流动法来测定吸附气体量的多少,而 BET 法既可以采用连续流动法,也可以采用容量法来测定吸附气体量。

Experiment 15 Determination of Solid Specific Surface Area by Solution Adsorption Method

【Purpose of the Experiment】

1. Master the methylene blue solution adsorption method to determine the specific surface area of activated carbon;

2. Understand the Langmuir monolayer adsorption theory and the basic principle of the solution adsorption method to determine specific surface area.

【Preview Requirements】

1. Understand the concept of specific surface;
2. Understand the Langmuir monolayer adsorption theory;
3. Understand the basic principle of measuring specific surface area by solution method.

【Principle】

There are many methods to determine the specific surface area of solids, including BET cryogenic adsorption method, gas chromatography, electron microscopy and so on. These methods require more complicated equipment or longer experimental time. In contrast, the solution adsorption method for the determination of specific surface area has the advantages of simple equipment and convenient operation.

The adsorption of water-soluble dyes can be used to determine the specific surface area of solids. Among all dyes, methine blue is a water-soluble dye, which can be easily adsorbed by solids. It has been reported that the adsorption of methine blue by most solids is monolayer adsorption within a certain concentration range, which is in line with Langmuir monolayer adsorption theory.

The basic assumption of Langmuir monolayer adsorption theory is shown as follows: the solid surface is uniform; the adsorption is monolayer; once the adsorbent is covered by the adsorbate, it cannot be re-adsorbed; adsorption and desorption establish a dynamic equilibrium; Before equilibrium, the adsorption rate is proportional to the unadsorbed surface, and the desorption rate is proportional to the coverage.

Presume the total number of adsorption sites on the solid surface is N, the coverage is θ, the concentration of adsorbate in solution is C. According to the above assumptions,

The adsorption rate is: $r_{adsorption} = k_1 N(1-\theta)C$ (k_1 is the adsorption rate constant)

The desorption rate is: $r_{desorption} = k_{-1} N\theta$ (k_{-1} is the desorption rate constant)

When the adsorption equilibrium is reached, the adsorption rate is equal to the desorption rate. That is,

$$k_1 N(1-\theta)C = k_{-1} N\theta$$

Therefore,

$$\theta = \frac{K_{adsorption} C}{1 + K_{adsorption} C} \quad (3-15-1)$$

In the above formula, $K_{adsorption} = k_1/k_{-1}$ is called adsorption equilibrium constant, and its value depends on the nature of adsorbent and adsorbate and temperature. The larger the $K_{absorption}$ value is, the stronger the adsorption capacity of solid is. If Γ represents the equilibrium adsorption capacity, and Γ_∞ represents the adsorption capacity when all adsorption sites are occupied, that is, the saturated adsorption capacity. Put $\theta = \Gamma/\Gamma_\infty$ into the formula of (3-15-1),

$$\Gamma = \Gamma_\infty \frac{K_{adsorption} C}{1 + K_{adsorption} C} \quad (3-15-2)$$

Therefore,

$$\frac{C}{\Gamma} = \frac{1}{\Gamma_\infty K_{adsorption}} + \frac{1}{\Gamma} C \quad (3-15-3)$$

The adsorption capacity can be calculated by the equilibrium concentration C and initial concentration C_0. The adsorption capacity Γ is calculated based on following formula,

$$\Gamma = \frac{(C_0 - C)V}{m} \quad (3-15-4)$$

In the formula, $V(L)$ is the total volume of the adsorption solution, and $m(g)$ is the mass of the adsorbent added to the solution. By calculating the value of Γ and plotting with $C/\Gamma - C$, Γ_∞ can be obtained from the slope of line, and $K_{absorption}$ can be obtained from the intercept. Γ_∞ refers to the saturated adsorption capacity of adsorbate by using per gram of adsorbent (expressed by the amount of substance). If the area occupied by each adsorbate molecule on the adsorbent is σ_A, the specific surface area of adsorbent can be calculated according to the following formula,

$$S = \Gamma_\infty L \sigma_A \quad (3-15-5)$$

In the formula, S is the specific surface area of adsorbent, and L is the Avogadro constant.

The structure of methine blue is

$$\left[\begin{matrix} H_3C \\ H_3C \end{matrix}\!\!>\!\!N\!\!-\!\!\underset{S}{\overset{N}{\bigcirc\!\!\bigcirc}}\!\!-\!\!N\!\!<\!\!\begin{matrix} CH_3 \\ CH_3 \end{matrix}\right]^{+}Cl^{-}$$

The cation size is $17.0 \times 7.6 \times 3.25 \times 10^{-30}$ m^3.

There are three orientations for adsorption of methine blue: the planar adsorption projected area is 135×10^{-20} m^2, the side adsorption projected area is 75×10^{-20} m^2, and the end group adsorption projected area is 39×10^{-20} m^2. For non-graphite activated carbon, methine blue is oriented by end group adsorption on the surface of activated carbon, so $\sigma_A = 39 \times 10^{-20}$ m^2. It can be seen that for solution adsorption, the orientation of non-spherical adsorbates is not completely same when they are adsorbed on the surface of various adsorbents, and the projected area of each adsorbate molecule can be very different, especially for the three adsorption orientations. There will be a big error between the calculated result and the real specific surface area. The measurement error of the solution adsorption method is usually 10% or even higher. More accurate measurement of specific surface area requires other methods.

According to the law of light absorption, when the incident light is monochromatic light, the absorbance of the solution is proportional to the concentration of the colored substance in solution and the thickness of solution layer, that is,

$$A = -\lg(I/I_0) = \varepsilon bC \qquad (3-15-6)$$

In the formula, A is the absorbance, I_0 is the incident light intensity, I is the transmitted light intensity, ε is the absorption coefficient, b is the optical path length or liquid layer thickness, and C is the solution concentration.

Methylene blue solution has two absorption peaks in the visible region: 445 nm and 665 nm. However, the adsorption of activated carbon at 445 nm has a great interference to the absorption peak, so the working wavelength of this experiment is 665 nm, and it is measured with a spectrophotometer.

【Apparatus and Reagents】

Instrument: 722 spectrophotometer and its accessories; volumetric flasks (500 mL); volumetric flasks (100 mL); volumetric flasks (50 mL); HY oscillator; sand core funnels; conical flask; droppers; electronic balance; muffle furnace.

Reagents: Methylene blue solution (about 0.2% original solution), methylene blue standard solution (0.3126×10^{-3} mol·L^{-1}), granular non-graphite activated carbon.

【Procedures】

1. Sample activation

Place the granular activated carbon in a porcelain worm and put it in a 500 ℃ muffle furnace for activation for 1 h, and then place it in a desiccator for later use. (This step has been done by the laboratory before the experiment.)

2. Solution adsorption

Take five dry conical flask with stoppers (numbered 1-5), then 0.1g of activated activated carbon is added into the bottles. Use a pipette to measure 30, 20, 15, 10 and 5 mL of the original methine blue solution with a concentration of about 0.2% into five 50 mL volumetric flasks, and dilute to the mark with distilled water. Pour the diluted methine blue solution into the five conical flasks with stoppers, and place them on the oscillator for 3 hours. After the sample is shaken to reach equilibrium, the filtrate after adsorption equilibrium can be obtained by filtration. Measure 5ml of the filtrate into a 500 mL volumetric flask, and dilute to the mark with distilled water. This is a balanced diluent.

3. Treatment of original solution

In order to test the concentration of about 0.2% methine blue original solution, 2.5 mL of solution is measured into a 500 mL volumetric flask, then dilute it to the mark with distilled water for use. This is a dilution of the original solution.

4. Preparation of Methylene Blue Standard Solution

Measure 0.5, 1, 2, 4, 8 and 12 mL of standard solution with a concentration of 0.3126×10^{-3} mol·L^{-1} into a 100 mL volumetric flask, then dilute it to the mark with distilled water for use. And the obtained solutions with the concentrations of 0.005, 0.01, 0.02, 0.04, 0.08, 0.12 × (0.3126×10^{-3} mol·L^{-1}) are numbered as A1, A2, A3, A4, A5 and A6, respectively.

5. Select the working wavelength

For methine blue solution, the working wavelength is 665 nm. Because the wavelength scale of each spectrophotometer has a slight error, take the A4 standard solution in step 4 and measure its absorbance in the range of 600~700 nm. the maximum absorbance wavelength is the working wavelength.

6. Measure absorbance

Choose a cuvette with high light transmittance $T\%$ as reference. Because the methine is adsorptive, it should be determined in the order from dilute to concentrate. At working wavelength, the absorbance of the standard solution A1 - A6 in step 4 and the methine blue solution prepared in step 2 and step 3 are tested, respectively.

【Experiment records and data processing】

1. Make a working curve of absorbance versus the concentration of methine blue solution.
2. Calculate the original and each equilibrium concentration of methine blue.

According to the absorbance of the original solution after dilution, check the corresponding concentration can be obtained from the working curve, which is multiplied with corresponding dilution factor in order to obtain the concentration C_0 of the original solution.

The absorbance of each diluted equilibrium solution measured in the experiment is checked from the working curve and the corresponding concentration is multiplied by the dilution factor to obtain the concentration C of the equilibrium solution.

3. Calculate the adsorption capacity. the adsorption capacity Γ can be calculated from the equilibrium concentration C and the initial concentration C_0 according to the formula (3 - 15 - 4). The $\Gamma - C$ adsorption isotherm curve can be obtained by Using the Γ.

4. Calculate the saturated adsorption capacity. Calculate the C/Γ value from the Γ and C data, then make the $C/\Gamma - C$ curves, and calculate the saturated adsorption capacity Γ_∞ from the curve. Use the dotted line to make a horizontal line on the $\Gamma - C$ chart. This dashed line is the asymptotic line of the adsorption capacity Γ.

5. Calculate the specific surface area of sample, and find the average value.

【Discussion】

1. The activated carbon particles should be uniform and dried.

2. The concentration of the standard solution must be accurately prepared.

3. The oscillation time should be sufficient, and the time is usually above 3 h in order to reach saturation.

4. According to the requirements of Langmuir adsorption isotherm, solution adsorption must be carried out under isothermal conditions, and the adsorbed conical flask is placed in a constant temperature water bath to achieve equilibrium. However, the room temperature change is not significant, so this experiment only placed the adsorption bottle on the shaker at room temperature. If the room temperature changes too much during the experiment, it will inevitably affect the results.

【Post Lab Questions】

1. Can the corresponding adsorption isotherm curve be drawn based on the adsorption capacity measured in this experiment?

2. When the solution is adsorbed, how to judge that it has reached equilibrium?

3. Why the solution should be diluted when determining the original solution and balance solution of methine blue?

实验 16 蛋白质等电点的测定

【实验目的】

1. 理解两性电解质的等电点、溶胶体系的渗透压和 Donnan 平衡等概念;
2. 学会用简单的方法测定明胶的等电点;
3. 理解 pH 值对干明胶吸水膨胀的影响。

【预习要求】

1. 了解蛋白质等两性电解质的性质;
2. 了解溶胶体系的渗透压和 Donnan 平衡等概念;
3. 了解 pH 计的使用方法及本实验的注意事项。

【实验原理】

蛋白质是两性化合物,由多种氨基酸组成。氨基酸分子中的氨基和羧基在水中都可解离。在酸性介质中,蛋白质分子以带正电的 NH_3^+RCOOH 形式存在;在碱性介质中,则以带负电的 NH_2RCOO^- 形式存在。当溶液中蛋白质分子上正、负电荷相等时,即它们不带正电也不带负电,这时溶液的 pH 值称为蛋白质的等电点。蛋白质的等电点由蛋白质的本性决定。换言之,等电点的 pH 值由蛋白质分子中氨基、羧基及水的解离常数决定。蛋白质随 pH 值改变进行的离子变换如下所示:

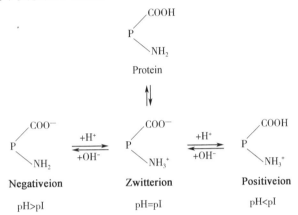

其中,pI 指蛋白质的等电点。

蛋白质分子的带电和水化是其在水中有一定稳定性的两个原因。由于在等电点时,蛋白质分子不带电,其水化程度小,因此,当加入有去水作用的物质(如无水乙醇等)时,蛋白质在越接近等电点的介质中越容易凝结。当溶液与纯溶剂(或浓度不同的另一溶液)被只允许溶剂分子通过的半透膜隔开时,溶剂分子将通过半透膜向溶液中扩散。为阻止溶剂的扩散所必须施加的阻力称为渗透压。或者说渗透压是溶剂分子通过半透膜扩散入溶液中的能力。亲液的高分子化合物渗透压相当大。渗透压的大小与溶质的分子量和浓度有关。

对于能电离的大分子物质(如蛋白质等)的溶液,当它与溶剂用半透膜隔开时,虽然膜内的小分子可以透过半透膜,但它在膜两边的分布并不均匀。这种因大离子需保持电中性的作用导致小离子分布不均匀的平衡称为 Donnan 平衡(唐南平衡)。当膜内小离子浓度远大于膜外小离子浓度时,膜外小离子几乎不向膜内扩散,而是溶剂向膜内扩散,从而形成附加的渗透压。

蛋白质种类较多,本实验以明胶为例。明胶是胶原的水解产物,是一种蛋白质,明胶软胶的网状结构可视为一种半透膜,溶剂和小离子可自由通过。明胶的等电点约为 pH=4.7。故当介质的 pH 值不在等电点时,明胶分子中的氨基或羧基发生解离,产生带解离氨基或羧基的大离子和与其相应的小离子。大离子不能透过软胶网状骨架所成的半透膜,小离子可透过。根据 Donnan 平衡产生附加渗透压,水分子将从凝胶外部向内部渗透,充填软胶的网孔,使明胶软胶体积膨胀。当介质的 pH 值等于等电点时,明胶分子不解离,不能形成附加渗透压,明胶软胶的体积无大变化。

本实验判断明胶等电点的方法是:向不同 pH 值的缓冲溶液中加入一定量的明胶液,再加入固定体积的无水乙醇,观察溶液浑浊程度(即明胶凝结程度)。此外,还在不同 pH 值下制备相同浓度的明胶软胶,观测其在水中的膨胀程度,以确定 pH 值对明胶软胶吸水膨胀的影响。

【仪器与试剂】

仪器:pH 计 1 台,恒温水浴 1 套,500 mL 和 100 mL 烧杯各 2 个,试管 9 支,量筒,培养皿。
药品:明胶、醋酸、醋酸钠、无水乙醇、盐酸、氢氧化钠。

【实验步骤】

1. 明胶等电点的测定

(1) 按表 3-16-1 在烧杯中配制不同 pH 值的缓冲溶液,各溶液用 pH 计测定其准确的 pH 值(也可按缓冲溶液理论计算 pH 值,计算值与测定值应十分接近);

(2) 在 9 支干燥的试管中,用移液管各取上述缓冲溶液 3 mL 并准确加入 1% 明胶液 1 mL,摇匀;

(3) 取出第 5 支试管(pH 值应为 4.75)向其中滴加无水乙醇,边加边摇,至溶液明显浑浊为止,记下加入乙醇的体积 V;

(4) 向其余 8 支试管中各滴加 V mL 的无水乙醇,边加边摇;

(5) 比较 9 支试管中溶液的浑浊程度,记下结果(浑浊程度可用不同数目的"—"符号表示),判断明胶的等电点。

表 3-16-1 明胶等电点测定数据

试剂	用量 V/mL								
	1	2	3	4	5	6	7	8	9
0.1 mol·L^{-1} NaAc	5.00	5.00	5.00	5.00	5.00	5.00	5.00	5.00	5.00
0.1 mol·L^{-1} HAc	0.31	0.62	1.25	2.50	5.00	10.00			
1.0 mol·L^{-1} HAc							2.00	4.00	8.00
水体积 / mL	14.69	14.38	13.75	12.50	10.00	5.00	13.00	11.00	7.00
总体积 / mL	20.00	20.00	20.00	20.00	20.00	20.00	20.00	20.00	20.00
测定的 pH 值									

2. 明胶的吸水膨胀

(1) 称取 30 g 明胶放入烧杯中,使其溶解在 270 mL 水中(适当加热),得 10% 明胶液。另配制 0.33 mol·L^{-1} 的 HCl 水溶液和 1.0 mol·L^{-1} 的 NaOH 水溶液备用;

(2) 在 9 个 100 mL 烧杯中,用量筒各加入 10% 明胶液 30 mL,再按表 2-16-1 的量加入所需的 HCl 溶液和 NaOH 溶液,补加水,使总体积为 40 mL,摇匀。趁热用 pH 计测定各液的 pH 值(若已凝结,可在水浴上稍微加热,使其熔化);

(3) 将配制好的不同 pH 值的明胶液分别倒入直径 8 cm 的培养皿中,置于冰箱中冷冻或自然凝结,制得弹性软胶;

(4)将明胶软胶切成边长1 cm的方块,将较为整齐的软胶块拨入已称重的100 mL烧杯中,再称重(凝胶量就不少于25 g);

(5)向已装入软胶的各烧杯中注满水,用细搅拌棒轻轻搅动软胶,不使胶块严重粘连,室温下(室温高时可置于冰箱中)放置2~4 h。

(6)用一表面皿盖住烧杯,将水尽可能地沥出(可控制沥水时间相近),称重,计算吸水量。

【实验记录及数据处理】

1. 以表格形式表示实验结果,判断明胶的等电点。

将以上实验结果一并记入表3-16-2中,以吸水量对pH值作图,总结出pH值对明胶吸水膨胀的影响。吸水量最小处对应的pH值即为明胶的等电点。

2. 解释pH值对明胶软胶吸水膨胀的影响。

表3-16-2 明胶软胶吸水膨胀数据

溶液	体积 V/mL								
	1	2	3	4	5	6	7	8	9
10%明胶									
0.33 mol·L^{-1} HCl									
1.0 mol·L^{-1} NaOH									
水体积/mL									
测定的pH值									
项目	数据								
空烧杯质量/g									
烧杯加软胶质量/g									
软胶质量/g									
烧杯加吸水后软胶质量/g									
总吸水质量/g									
单位质量软胶吸水量/(g·g^{-1})									

【讨论与说明】

1. 等电点测定的实验要求各种试剂的浓度和加入量必须相当准确;

2. 缓冲液的pH值必须准确。

【思考题】

1. 什么是蛋白质的等电点?

2. 在等电点时,蛋白质溶液为什么容易发生沉淀?
3. 本实验也可以鲜鸡蛋为材料来开展。

【相关阅读】

等电聚焦电泳法测定蛋白质等电点

等电点是蛋白质组分的特性量度,不同蛋白质各有特异的等电点,在等电点时,蛋白质的理化性质都有变化,可利用此种性质的变化测定各种蛋白质的等电点。最常用的方法是测其溶解度最低时的溶液 pH 值,如本实验所述。

在水溶液中的蛋白质分子由于表面生成水化层和双电层而成为稳定的亲水胶体颗粒,在一定的理化因素影响下,蛋白质颗粒可因失去电荷和脱水而沉淀。蛋白质的沉淀反应可分为两类。

(1) 可逆的沉淀反应:此时蛋白质分子的结构尚未发生显著变化,除去引起沉淀的因素后,蛋白质的沉淀仍能溶解于原来溶剂中,并保持其天然性质而不变性。如大多数蛋白质的盐析作用或在低温下用乙醇(或丙酮)短时间作用于蛋白质。提纯蛋白质时,常利用此类反应。

(2) 不可逆沉淀反应:此时蛋白质分子内部结构发生重大改变,蛋白质因变性而沉淀,不再溶于原来溶剂中。加热引起的蛋白质沉淀与凝固,以及蛋白质与重金属离子或某些有机酸的反应都属于此类。

蛋白质变性后,有时由于维持溶液稳定的条件仍然存在(如电荷),并不析出,因此,变性蛋白质并不一定都表现为沉淀,而沉淀的蛋白质也未必都已变性。

目前,等电聚焦电泳法是测定蛋白质等电点的常用方法。等电聚焦又称电聚焦、聚焦电泳等。

蛋白质(酶)、多肽等两性电解质,其所带电荷的性质和数量随所处环境的 pH 值而变化。环境 pH 值低于其等电点时带正电,在电场中向阴极移动;环境 pH 值高于其等电点时带负电,在电场中向阳极移动;当环境的 pH 值等于其等电点时则不带电荷,在电场中不移动。据此,在电泳系统中创造了一个由阳极至阴极,pH 值由低到高(即环境由酸变碱)的连续而稳定 pH 值梯度环境,那么处在这种系统中具有不同等电点的各种蛋白质,将据所处环境的 pH 值与其自身等电点的差别,分别带上正电荷或负电荷,并向与它们各自的等电点相当的 pH 值环境位置处移动,当到达该位置时即停止移动,从而各自焦聚(聚焦),分别形成一条集中的蛋白质区带。这种根据蛋白质等电点的不同而将它们分离开的电泳方法称为等电聚焦电泳。电泳后测定各种蛋白质聚焦部位的 pH 值,即可得知它们的等电点。

在等电聚焦电泳中,造成环境由酸至碱逐步变化所用的物质是一类两性电解质。它具有依次递变但又相差不大的等电点(pI),在电场中可以形成逐渐递变而又连续的 pH 值梯度。此类物质在等电点处具有足够的缓冲能力,以保证不受蛋白质样品等两性物质对 pH 值梯度的影响;在等电点处还具有足够高的电导,保证电流通过,而且整个体系的电导均匀,使样品迁移不受影响,达到聚焦;这类物质不与分离物质反应或使之变性,自身化学性质不同于被分离物质,分子量也小,当电泳后经透析可与被分离物质分开。

1969 年,瑞典 Vesterberg 首先合成这种物质,随即由瑞典 LKB 公司以 Ampholine 为商品名进行销售。Ampholine 由多乙烯多胺(如三乙烯四胺、五乙烯六胺等)与丙烯酸(不饱和酸)进行加成反应而生成,是一系列含不同比例氨基和羧基的氨基羧酸的混合物。其为无色水溶液,含量为40%,分子量300~1 000。它的水溶性很好,1%水溶液中的紫外吸收(260 nm)很低。商品 Ampholine 的 pH 值有各种范围,最宽的 pH 值为3~10。测定未知样品的等电点时,首先用 Ampholine 进行初步测定,根据测得的结果,再选择合适的 pH 值范围较窄的 Ampholine 进行精确测定。

为防止电泳过程中,因为对流现象使已经聚焦的蛋白质区带再混合,需要一种能抗对流的介质作为电泳支持物。目前常用的有蔗糖、甘油或乙二醇形成的密度梯度;用琼脂糖、聚丙烯酰胺形成的凝胶及利用亲水惰性颗粒,如葡聚糖凝胶、淀粉、滤纸等为支持剂,以凝胶作支持物的方法称为凝胶聚焦电泳,这种方法适用于分析分离,其他支持剂适用于制备。

等电聚焦电泳的优点如下:
(1) 分辨率高,可将等电点相差0.01~0.02 pH 单位的蛋白质分开;
(2) 不像一般电泳易受扩散作用影响,使区带越走越宽;聚焦电泳能抵消扩散作用,使区带越走越窄;
(3) 由于等电聚焦作用,很稀的样品也可以聚焦而浓缩;
(4) 因为是根据等电点特性分离,所以重复性好,精确度高,可达0.01 pH 单位。

其不足之处如下:
(1) 要求样品溶液无盐,因为盐会增大电流量,产生热量;盐分子移至两极时,将产生酸或碱,中和两性电解质;
(2) 要求样品在等电点时稳定,不适宜用于在等电点时不溶解或变性的蛋白质。

等电聚焦电泳具有分离、制备及鉴定蛋白质、多肽的多种功能。

Experiment 16　Determination of Protein Isoelectric Point

【Purpose of the Experiment】

1. Understand concepts such as the isoelectric point of the ampholyte, the osmotic pressure of the sol system and the Donnan equilibrium
2. Learn to use a simple method to determine the isoelectric point of gelatin
3. Understand the influence of pH on water swelling of dry gelatin

【Preview Requirements】

1. Understand the properties of ampholytes such as protein
2. Understand concepts such as osmotic pressure and Donnan balance of the sol system
3. Understand the usage method of pH meter and the precautions of this experiment.

【Principle】

Protein is an amphoteric compound, it is composed of a variety of amino acids. Both the amino group and the carboxyl group in the amino acid molecule can be dissociated in water. In acidic medium, protein molecules exist in the form of positively charged NH_3^+RCOOH; in alkaline medium, they exist in the form of negatively charged NH_2RCOO^-. When the positive and negative charges on the protein molecules in the solution are equal, that is, they are neither positive nor negative, the pH value of the solution at this time is called the isoelectric point of the protein. The isoelectric point of a protein is determined by the nature of the protein. In other words, the pH value of the isoelectric point is determined by the dissociation constant of the amino group, the carboxyl group and the water in the protein molecule. The ion transformation of protein with pH changes is shown in the following formula: Among them, pI refers to the isoelectric point of the protein.

The charging and hydration of protein molecules are two reasons for their stability in water. Because of the isoelectric point, the protein molecule is not charged, its degree of hydration is reduced. Therefore, when adding substances that have the effect of dewatering, the closer the protein is to the isoelectric point, the easier it is to coagulate. When the solution and the pure solvent are separated by a semi-permeable membrane that only allows solvent molecules to pass through, the solvent molecules will diffuse into the solution through the semi-permeable membrane. The resistance that must be applied to prevent the diffusion of the solvent is called osmotic pressure. Osmotic pressure is the ability of solvent molecules to diffuse into the solution through the semi-permeable membrane. The osmotic pressure of the lyophilic polymer compound is quite high. The value of the osmotic pressure is related to the molecular weight and concentration of the solute.

$$P\begin{cases}COOH\\NH_2\end{cases}$$
Protein

$\updownarrow$

$$P\begin{cases}COO^-\\NH_2\end{cases} \underset{+OH^-}{\overset{+H^+}{\rightleftharpoons}} P\begin{cases}COO^-\\NH_3^+\end{cases} \underset{+OH^-}{\overset{+H^+}{\rightleftharpoons}} P\begin{cases}COOH\\NH_3^+\end{cases}$$

Negativeion   Zwitterion   Positiveion

pH>pI   pH=pI   pH<pI

For the solution of ionizable macromolecular substances (such as proteins, etc.), when it is separated from the solvent by a semi-permeable membrane, the small molecules in the membrane can penetrate the semi-permeable membrane, its distribution on both sides of the membrane is not uniform. This kind of balance in which small ions are distributed unevenly due to the need to maintain electrical neutrality of large ions is called Donnan balance. When the concentration of small ions inside the membrane is much more than the concentration of small ions outside the membrane, small ions outside the membrane hardly diffuse into the membrane, but the solvent diffuses into the membrane, thereby forming an additional osmotic pressure.

There are many types of protein, this experiment takes gelatin as an example. Gelatin is a hydrolysate of collagen, it is a protein, the network structure of soft gelatin can be regarded as a semi-permeable membrane, solvents and small ions can pass freely. The isoelectric point of gelatin is about pH 4.7. Therefore, when the pH value of the medium is not at the isoelectric point, the amino groups or carboxyl groups in the gelatin molecule dissociate, resulting in large ions with dissociated amino groups or carboxyl groups and corresponding small ions. Large ions cannot pass through the semi-permeable membrane formed by the soft rubber mesh frame, while small ions can pass through. According to Donnan's equilibrium, additional osmotic pressure is generated, water molecules will penetrate from the outside to the inside of the gel, filling the mesh of the soft gel, and expanding the volume of the gelatin soft. When the pH value of the medium is equal to the isoelectric point, the gelatin molecule does not dissociate and cannot form additional osmotic pressure, and the volume of the soft gelatin does not change much.

In this experiment, the method to determine the isoelectric point of gelatin is: add a certain amount of gelatin to buffer solutions of different pH values, and then add a fixed volume of absolute ethanol to observe the degree of turbidity of the solution (the degree of gelatin condensation). In addition, the same concentration of soft gelatin was prepared under different pH values, and the degree of swelling in water was observed to determine the effect of pH on the swelling of soft gelatin.

【Apparatus and Reagents】

Apparatus: pH meter; Constant temperature water bath; 500 mL beakers; 100 mL beakers; Test tubes; Measuring cylinders; Petri dishs.

Reagent: Gelatin; Acetic acid; Sodium acetate; Absolute ethanol; Hydrochloric acid; Sodium hydroxide.

【Procedures】

1. Determination of the isoelectric point of gelatin

A. Prepare buffer solutions of different pH values in a beaker according to Table 3-16-1, and measure the accurate pH value of each solution with a pH meter;

B. In nine dry test tubes, use a pipette to take the above buffer solution of 3 mL and accurately add 1% gelatin solution of 1 mL to shake well;

C. Take out the fifth test tube (pH value should be 4.75) and add absolute ethanol dropwise to it, and shake it while adding it until the solution is obviously muddy until it becomes turbid, write down the volume V of ethanol added;

D. Add V mL absolute ethanol dropwise to each of the remaining 8 test tubes, and shake while adding;

E. Compare the degree of turbidity of the solution in the 9 test tubes, write down the results (the degree of turbidity can be indicated by a different number of "-" symbols), and determine the isoelectric point of the gelatin.

Table 3-16-1 Gelatin isoelectric point measurement data

Reagent	V/mL								
	1	2	3	4	5	6	7	8	9
0.1 mol·L^{-1} NaAc	5.00	5.00	5.00	5.00	5.00	5.00	5.00	5.00	5.00
0.1 mol·L^{-1} HAc	0.31	0.62	1.25	2.50	5.00	10.00			
1.0 mol·L^{-1} HAc							2.00	4.00	8.00
Water/mL	14.69	14.38	13.75	12.50	10.00	5.00	13.00	11.00	7.00
Total Volume/mL	20.00	20.00	20.00	20.00	20.00	20.00	20.00	20.00	20.00
Measured pH									

2. Water swelling of gelatin

A. Weigh 30 g of gelatin and put it in a beaker and dissolve it in water of 270 mL (appropriately heated) to obtain a 10% gelatin solution. Prepare 0.33 mol/L HCl aqueous solution and 1.0 mol/L NaOH aqueous solution for later use;

B. In nine beakers of 100 ml, add 10% gelatin solution of 30 ml to each measuring cylinder, then add the required HCl solution and NaOH solution according to the amount in Table 3-16-1, and add water to make the total volume 40 mL, shake well. Measure the pH value of each liquid with a pH meter while it is hot (if it has condensed, heat it slightly on a water bath to melt it);

C. Pour the prepared gelatin solutions of different pH values into petri dishes with a diameter of 8 cm, and place them in the refrigerator to freeze or coagulate naturally to obtain elastic soft gels;

D. Cut the soft gelatin gel into squares with a side length of 1 cm, transfer the neat soft gel block into a weighed 100 mL beaker, and then weigh it (the amount of gel should not be less than 25 g);

E. Fill each beaker filled with soft glue with water, and gently stir the soft glue with a fine stirring rod to prevent the glue blocks from sticking seriously. Place it at room temperature for 2~4 h. (it can be placed in the refrigerator when the room temperature is high);

F. Cover the beaker with a watch glass, drain the water as much as possible (the draining time can be controlled to be similar), weigh and calculate the water absorption.

【Data Records and Analysis】

1. Show the experimental results in tabular form to determine the isoelectric point of gelatin. Record the above experimental results in Table 3-16-2, plot the water absorption versus pH value, and summarize the effect of pH value on the water swelling of gelatin. The pH value corresponding to the minimum water absorption is the isoelectric point of gelatin.

2. Explain the effect of pH on the water swelling of soft gelatin.

Table 3-16-2 Water swelling data of soft gelatin

Solution	V/mL								
	1	2	3	4	5	6	7	8	9
10% gelatin									
0.33 mol/L HCl									
1.0 mol/L NaOH									
Water volume/mL									
Measured pH									
	Data								
Empty beaker quality/g									
Beaker plus sof glue quality/g									
Soft rubber quality/g									
After adding water to the beaker Soft glue quality/g									
Total water absorption quality/g									
Unit mass soft glue Water absorption/g									

【Notes】

1. The experiment of isoelectric point determination requires that the concentration and amount of various reagents must be quite accurate.

2. The pH value of the buffer solution must be accurate.

【Post Lab Questions】

1. What is the isoelectric point of protein?

2. At the isoelectric point, why is the protein solution prone to precipitation?

3. This experiment can also be carried out with fresh eggs.

第4章 设计型实验

实验17 电还原草酸制备乙醛酸的方法

【实验目的】

1. 了解电化学在有机电合成中的应用;
2. 利用本实验提供的设备,设计、实施电还原草酸制备乙醛酸的实验,从而掌握直接电合成的一般方法;
3. 了解电合成实验结果的评价方法。

【预习要求】

1. 查阅文献,设计电还原草酸制备乙醛酸的实验方案;
2. 查阅文献,确定草酸及乙醛酸的分析方法;
3. 查阅文献,确定对实验结果进行评价的方法。

【实验原理】

许多合成有机化合物的反应中包含着电子的转移,因此可以安排在电池中进行,这就是有机电合成反应。研究用电化学方法进行有机化合物合成的技术叫作有机电合成。有机电合成具有以下优点:①洁净,以电子的得失完成了氧化还原反应,不需要外加氧化剂和还原剂;②条件温和,如在常温、常压下即可完成有机合成,尤其对不稳定的复杂分子结构的有机物的合成尤为有利;③副产物少;④节能,一方面体现在综合能耗上,另一方面是由于极间电压低(2~5 V),可接近热力学的要求值;⑤易控,反应速度完全可以通过调节电流来实现,为自动化连续操作打下了基础;⑥规模效应小,对精细化工产品的生产尤为有利。由于有机电合成具有以上优点,它基本上符合"原子经济性"的要求,因此必将成为21世纪各化学基础学科和应用技术研究的热点。

基本的有机电合成可分为直接电合成、间接电合成、配对电合成及自发电合成。直接电合成是指原料直接在电极上发生电极反应转化为合成产物,反应直接在电极表面完成。间接电合成是指选择某种氧化还原对作为"媒质",这种媒质能在电极表面上首先被氧化或还原,然后在与有机原料反应合成目的产物,分离出产物后,媒质可以在电解槽中通过阳极或阴极反应再生,循环使用。间接电合成法可按两种方式操作:槽内式和槽外式。有机电合成反应只发生在某一极,而另一极上发生的反应没有被利用,如果在阴阳两极上进行生成目的产物的反应,就称为配对电合成。多数有机电合成都是通过电解过程来实现的,但有的有机化合物的生成反应的 $\Delta G < 0$,因此可以安排在电池中进行,在合成目的产物的同时还可以获

得电能,这种称为自发电合成。本实验拟采用直接电合成技术,在阴极电还原草酸制备乙醛酸。

乙醛酸又名二羟醋酸、甲醛甲酸,是一种最简单的醛酸,兼有酸和醛的性质,是有机合成的重要原料,被广泛用于香料、医药和农药等精细化学品的生产。其制备方法有乙二醛硝酸氧化法、顺丁烯二酸臭氧氧化法、乙二醛电化学氧化法、草酸电化学还原法等。与其他方法相比,草酸电化学还原法具有原料便宜且来源充足、工艺路线短、工艺条件温和、污染小等优点,因此是一种很有潜力的生产方法。

草酸电还原制备乙醛酸的电极反应为

阳极: $H_2O = 2H^+ + \frac{1}{2}O_2\uparrow + 2e$

阴极: $COOHCOOH + 2H^+ + 2e = CHOCOOH + H_2O$

阴极副反应: $COOHCOOH + 2H^+ + 2e = CH_2OHCOOH + H_2O$

$2H^+ + 2e = H_2\uparrow$

采用合适的电极材料及电解条件可抑制副反应,尽可能多地制取目的产物乙醛酸。电解条件包括温度、槽电压、电流密度、阴阳极面积比等。

评价有机电合成的实验结果的指标有电流效率 η_i、电压效率 η_V、电能效率 η_W、转化率 α、产率 Y、时空产率 Y_{ST} 等,其中最常用的是电流效率 η_i、转化率 α、产率 Y、时空产率 Y_{ST}。

(1) 电流效率 $\eta_i = \dfrac{Q_t}{Q_P} \times 100\% = \dfrac{m_t}{m_P} \times 100\%$

其中,Q_t 和 Q_P 分别为生成一定量产物所消耗的理论电量和实际电量,C;m_t 和 m_P 分别为通过一定电量理论上应生成的产物量和实际生成的产物量,kg。

(2) 电压效率 $\eta_V = \dfrac{E_t}{E_P} \times 100\%$

其中,E_t 和 E_P 分别为理论分散电压和实际槽电压,V。

(3) 电能效率 $\eta_W = \dfrac{W_t}{W_P} \times 100\%$

其中,W_t 和 W_P 分别为理论功和实际功。

(4) 转化率 $\alpha = \dfrac{\Delta m}{m_0} \times 100\%$

其中,Δm 为转化的主反应物量,kg;m_0 为主反应物初始量,kg。

(5) 产率 Y $Y = \dfrac{\Delta m'}{\Delta m} \times 100\%$

其中,$\Delta m'$ 为转化产物消耗的反应物量,kg。

(6) 时空产率 Y_{ST} $Y_{ST} = \dfrac{m'}{tV} \times 100\%$

其中,m' 为生成目的产物的量,kg;t 为反应时间,h;V 为电化学反应器的体积,m^3。

【仪器与试剂】

仪器:H 形隔膜式电解槽 1 只,磁力搅拌器 1 台,直流稳压电源(30V,2A),电炉及电压表各 1 只,电流表 1 只,玻璃水浴缸 1 个,铅板 1 块,分析天平 1 台,螺丝,导线,电工工具,必要的玻璃仪器。

药品:化学纯草酸及分析用试剂。

【设计要求】

1. 查阅文献,根据本实验所提供的仪器设计出电还原草酸制备乙醛酸的实验方案,并进行实验。

2. 查阅文献,确定草酸及乙醛酸的分析方法。

3. 根据实验及分析结果用电流效率、产率及转化率对实验结果进行评价。

【思考题】

1. 影响电合成的因素主要有哪些?在研究中通常如何考查这些因素的影响?
2. 优选工艺条件时主要参照哪些指标?
3. 分析电解液中乙醛酸和草酸含量时应注意哪些问题?
4. 如何对电合成结果进行评价?

【相关阅读】

有机物的电化学合成

早在19世纪初期,雷诺尔德(Rheinold)和欧曼(Erman)发现电是一种强有力的氧化剂和还原剂,那时他们就已经用醇稀溶液进行过电解反应的研究。1934年,法拉第首先使用电化学法进行了有机物的合成和降解反应研究,发现在醋酸盐水溶液中电解时,阴极上会析出CO_2,并生成烃类化合物。后来,柯尔贝(Kolbe)在法拉第工作的基础上,创立了有机电化学合成(又称有机电解合成,以下简称有机电合成)的基本理论。

虽然有机电合成的研究早在19世纪初就已经开始,但是限于理论和工艺复杂性及有机催化合成迅速发展带来的竞争,有机电合成在很长一段时间内进展缓慢,只是作为有机化学家们在实验室中制备有机化合物的一种常用方法,并未在工业化上迈出步伐。直到20世纪50年代,电化学理论、技术、新材料的发展为有机电合成的工业应用奠定了基础。有机电合成真正取得实质性进展开始于1960年,美国孟山都(Monsanto)公司电解丙烯酸二聚体生产己二腈获得了成功,并建成年产1 145万吨的己二腈生产装置,这是有机电合成走向大规模工业化的重要转折点。从此,有机化合物的电化学性质和有机电化学反应机理的研究得到了快速发展,以有机电合成为基础的工业领域不断出现。

由于有机电合成具有污染小(甚至无污染)、产物收率和纯度高、工艺流程较短、反应条件温和等优点,近年来,世界工业先进国家有机电合成的发展非常迅速,目前已有上百种有机化工产品通过电化学合成实现了工业化生产或者进入了中试阶段。近年来每年发表的有关有机电化学合成方面的研究论文有几百篇,有关的专利发明每年平均有50~70项之多,这些数字表明有机电合成工业已引起人们的足够重视,并在高科技领域内崭露头角。

我国电合成方面的研究起步较晚。20世纪80年代以来,我国已有许多研究者涉足这一领域,做了大量研究开发工作,在有机电合成领域得到了较大的发展,有10多个产品实现了工业化,研究的品种也日趋增多。我国有机电合成科学和技术与世界的差距正在逐步缩小。

1. 有机电合成研究内容

(1) 电极过程动力学

电极过程动力学包括扩散动力学和电化学步骤动力学。扩散动力学和非均相化学反应中的扩散动力学没有明显区别,包括对流、扩散、电迁移等现象。电化学步骤动力学体现了电极过程的核心内容,包括化学反应和电子传递过程。研究电极过程首先要研究整个有机电极反应的基本历程,并弄清各步骤的动力学特征和机理,这是研究有机电合成问题的关键。为了达到这一目的,往往需要弄清下列三个方面的情况:①弄清整个电极反应的历程,即所研究的电极反应包括哪些步骤以及它们的组合顺序;②在组成电极反应的各步骤中,找出决定整个电极反应速度的控制步骤;③测定控制步骤的动力学参数(即整个电极反应的动力学参数)及其他步骤的热力学参数。

(2) 电极材料

电极既是电化学过程的催化剂,又是电极反应进行的场所,电极材料的性质对整个电合成反应途径和选择性都有很大的影响,因此有关电极材料的研究成为近些年来有机电合成研究的热点。电极材料的选择可以考虑以下原则:①导电性;②对过电位、耐腐蚀性、机械加工性能等方面的要求;③对电极的形状和结构的要求;④对电极表面的性质要求。常用的阴极材料有汞、铅、锡、铜、铁、铝、铂、镍和碳等。由于阳极材料在阳极反应中的腐蚀问题,合适的阳极材料是非常少的。实验室中常用的有铂、金和碳。在稀硫酸介质中,一般采用铅或铅银合金电极。用钛基或陶瓷基二氧化铅涂层电极可解决阳极的腐蚀问题,但涂层与基体的结合力较差,涂层易剥落造成电极失活,若在涂层与基体之间加上锡锑等中间层,可以改善涂层与基体的结合力。目前,二氧化铅电极的制备方向是将基体制成多孔电极,把二氧化铅以微粒的形式镶嵌在电极表面。这种电极不但涂层与基体的结合力好、寿命长,而且电极比表面积大、产率高。氯碱工业中使用的钌钛电极寿命可达 2~8 年,电流密度约为 2 000 A·m^{-2},但由于有机电解合成中许多反应均在硫酸溶液中进行,因此现有的钌钛电极显然不适合,故阳极材料仍是有机电解合成工业中一个亟待解决的关键问题。

(3) 离子交换膜

为了防止阴极或阳极产物进一步在阳极氧化或在阴极还原,需要用离子交换膜将阴、阳两室分开。离子交换膜的典型材质是全氟磺酸酯及全氟磺酸酯羧酸酯,以交联的接枝膜最为适宜。可以说,离子交换膜是有机电解合成工业中的又一技术关键问题。国内有机所、有机氟材料研究所、上海原子核研究所和华东理工大学等都在阳离子交换膜的工业化上做了大量工作,但还需要在降低成本、延长寿命、提高离子选择性透过率等方面做一些工作,以提高有机电合成相对于化学合成的竞争能力。

(4) 电化学反应器

电化学反应器可分为平板电极反应器、旋转圆柱电极反应器、固定床电极反应器和流化床电极反应器,前两种为二维电极反应器,后两种为三维电极反应器。每一种反应器又可以有不同的反应器结构形式。固定床电极反应器内电势、电流密度(反应速率)及流体分布是三维电化学反应器内的特有现象,电势及电流分布与电极的几何形状、几何尺寸、床层空隙率、电解液的电导率、流体力学性质、极化类型、极化程度和操作条件等因素有关。床层内的电势与电流分布对反应器的空速、反应选择性、单程转化率、电流效率等都有影响。近年来,随着四乙基铅、硝基苯电解还原制备对氨基苯酚和苯氧化制备对苯醌等一系列过程在固定床反应器中实现工业化,固定床电极反应器成了有机电合成工业中的首选反应器,对固定床电极反应器的基础理论也有了较深入的研究,并有较详尽的分析报道。

2. 有机电化学合成研究进展和有待解决的问题

有机电合成的研究进展迅速,研究领域包括直接电合成、间接电合成、界面修饰电极、反应性电极等。除此之外,在下述领域也取得了很大进展:①固体聚合物电解质(简称SPE)在电化学中的应用。SPE是一种高分子离子交换膜,由于其较好的化学和机械稳定性、优良的导电性等优点,目前逐渐应用于氯碱工业、电解水工业以及航空航天用燃料电池、核潜艇用氧气发生器等领域,使这些领域的技术水平取得了革命性的进步。②碳载Sb-Pb-Pt电催化纳米材料的最新研究进展。电极材料一直是电化学研究的重点,寻找和研制高活性、高选择性的新型电催化剂材料具有重要的意义。近年来,在导电载体上沉积纳米材料制备高性能实用型电催化剂引起了广泛的关注。实验表明,碳载Sb-Pb-Pt电催化纳米材料的催化活性和稳定性远高于常用的Sb和Pb等金属电极,应用前景很好。③金属有机物合成研究的最新进展。电合成金属有机物具有选择性高、产品纯度高、环境污染少等优点,因此具有很大优势。金属有机物具有特殊的功能,可用作催化剂、聚合材料、稳定剂、防腐剂和颜料等,近年来需求量增长很快。④超声在有机电合成中的最新应用。超声对有机电合成具有多种作用,超声的应用为解决电合成中的许多问题,特别是最佳电化学反应条件提供了途径,展示了良好的工业应用前景。因此,超声在有机电合成中的应用研究是当前有机电合成研究的前沿领域。此外,有机电合成在仿生合成、医药、信息产品、食品添加剂等精细有机化工产品的合成方面也取得了很多突破性进展。

有机电合成作为一门新的热点学科,其不足之处主要有以下几点:①电解反应仅限于氧化和还原反应。②反应装置比较复杂。由于存在"两极"的差别且两极分别有氧化产物和还原产物,再加上要保证反应物和目的产物的扩散分离,因此往往需要对电极材料、电解槽结构和隔膜材质提出很高的要求。此外,槽外设备也增加了电解装置的复杂性。③合成理论及工艺技术不够成熟,尤其是电合成反应动力学原理中许多问题有待深入研究。另外,在均匀分布、分离技术方面也存在难题。为了克服上述不足,进一步增强有机电合成的优势,今后必须在以下几个方面进行深入研究和开发:以固定床、流化床三维电极取代空间反应界面小的板式或网式二维电极,同时采用媒质反应技术和相转移催化技术;采用成对电解合成技术以期成倍地增加电流效率和电能效率;推进电化学工程的研究,使电解反应器的设计、控制以及电解槽的放大更趋合理可行,以降低成本;把注意力从产值低、数量大的电解合成产品转向产值高、数量小的精细化工产品。

实验18 电 镀 铜

【实验目的】

1. 掌握电镀和化学镀的原理;
2. 掌握电镀和化学镀的工艺流程;

3. 掌握影响镀层的因素;
4. 培养学生设计实验的能力。

【预习要求】

1. 查阅文献,设计出电镀铜的实验方案;
2. 查阅文献,确定对实验结果进行评价的方法。

【实验原理】

电镀就是利用电解的方式使金属或合金沉积在工件表面,以形成均匀、致密、结合力良好的金属层的过程。

电镀铜是使用最广泛的一种预镀层。锡焊件、铅锡合金、锌压铸件在镀镍、金、银之前都要镀铜,用于改善镀层结合力。铜镀层是重要的防护装饰性镀层铜/镍/铬体系的组成部分,柔韧而孔隙率低的铜镀层对于提高镀层间的结合力和耐蚀性起重要作用。目前,使用最多的镀铜溶液是氰化物镀液、硫酸盐镀液和焦磷酸盐镀液。碱性镀铜易获得较薄的、细致光滑的铜镀层,为了获得较厚的铜镀层,必须先将镀件进行碱性镀铜,再置于含有硫酸铜、硫酸镍和硫酸等成分的电解液中进行酸性镀铜。此外,还有焦磷酸盐、酒石酸盐、乙二胺等配制的无氰电解液。焦磷酸盐电解液已被广泛采用,而塑料电镀制品也越来越受到人们的青睐。塑料电镀产品的抗拉应力、扭应力和冲击应力都增大许多。塑料电镀制品能大大增强抗腐蚀能力,能增加塑料的散热能力,而且在质量上,塑料制品比其他以金属为基底的制品轻许多,塑料制品成本较金属制品低很多,生产效率也高。由于塑料电镀制品具有上述优点,故对塑料电镀方法的研究及生产工艺的开发具有很大的实际意义。

塑料电镀的方法有湿法电镀、真空喷镀(涂)及离子溅射镀等。其中,湿法电镀是最常见的电镀方式,是一个电化学的过程,利用正负电极,加以电流在渡槽中进行,可镀金、镀银、镀镍、镀铬、镀镉等,电镀液污染很大,但良好的结合力具有耐蚀和耐热性,因而得到广泛应用。涂镀是利用专门配置的两种涂液,在金属件需要镀的部分不停"涂刷",在涂刷区域产生化学反应,堆积出一个涂层,手工操作,用于工件上面的"加料",以达到尺寸要求,常用于柴油机曲轴、连杆等的处理。真空镀是把金属在高的真空度下用电极加热200 ℃左右,使得金属升华产生蒸气,附着在镀件的表面,形成镀层,常用来镀金属镉、银,几乎无污染。离子溅射镀工业化应用还比较少。

塑料电镀的工艺流程通常如下:

化学除油→流水洗→化学粗化→流水洗→敏化处理→流水洗→纯水洗→活化处理→流水洗→还原处理→化学镀铜→流水洗→光亮镀铜→流水洗→光亮镀镍→流水洗。

【仪器与试剂】

仪器:烧杯,玻棒,试剂瓶若干,ABS塑料,恒电位仪1台,导线若干,恒温水浴槽。

试剂:无水碳酸钠,磷酸三钠,硫酸,铬酐,氯化亚锡,盐酸,硝酸银,氨水,甲醛,酒石酸钾钠,氢氧化钠,硫酸铜,氯化镍,硫酸镍,硼酸(以上试剂均为分析纯)。

【设计要求】

1. 查阅文献,根据本实验所提供的仪器设计出电镀铜的实验方案,并进行实验。
2. 根据实验结果分析镀层的结合力及其外观光亮程度。

【思考题】

1. 如何对实验结果进行评价?
2. 哪些参数会对实验结果产生较大影响?
3. 电镀和化学镀有何区别?

【相关阅读】

1. 电 镀 技 术

电镀是利用电化学方法对金属或非金属制品进行表面加工,使它们在表面上获得各种不同的金属镀层,以提高金属表面的抗腐蚀性能和装饰性能,改善制品表面的机械性能,给制品以特殊的物理化学特性。金属镀层的应用十分广泛,已经广泛应用到国民经济各部门和多个科学领域。例如,汽车制造就需要把外露的零件进行电镀防止腐蚀,或在表面镀上光亮铬达到装饰的目的。大量轻工产品如自行车、缝纫机、钟表、灯具、照相机等,不但要求它们不锈蚀,而且要求能长期保持其漂亮的外观。电子元件需提高其表面的导电性能应镀银,提高焊接性能可镀锡或铅锡合金。塑料制品也可以进行电镀,它的应用范围更加广泛。除此之外,自20世纪末国内外开始推广了刷镀,也就是无槽电镀,解决了大工件难以电镀的困难,可以直接、快速地加工和现场修复,使电镀工艺在国民经济中有着更广泛的应用。

电镀根据电镀后产品的用途主要分为防腐蚀电镀、装饰性电镀、抗磨损性电镀及导电性电镀等,而根据电镀后外观的效果有高光电镀、亚光电镀、珍珠铬电镀、蚀纹电镀、混合电镀、局部电镀和彩色电镀等。高光电镀、亚光电镀及珍珠铬电镀的效果实现均首先要求模具表面良好抛光,然后对注射出的塑件分别采用光铬处理、亚铬处理及珍珠铬处理后得到效果。蚀纹电镀效果的实现通常要求模具表面处理出不同效果的蚀纹方式后,注射出的塑件采用光铬处理后得到效果。在模具处理上既有抛光的部分又有蚀纹的部分,注射出的塑件电镀后出现高光和蚀纹电镀的混合效果,突出某些局部的特征,属于混合电镀。而有时需要通过局部电镀使得成品件的表面局部没有电镀的效果,与有电镀的部分形成反差,形成独特的设计风格。此外还有彩色电镀,即采用不同的电镀溶液,使塑件表面沉积的金属反射出不同的光泽,形成独特的效果。

2. 镀铜工艺流程示例参考

(1) 化学镀铜

① 化学除油

配制 $0.20 \sim 0.30$ mol·L^{-1} 碳酸钠, $0.25 \sim 0.30$ mol·L^{-1} 磷酸三钠和 $45 \sim 50$ g·L^{-1} 洗涤液的混合水溶液,在 $40 \sim 50$ ℃ 温度下恒温清洗 $25 \sim 30$ min。

②化学粗化液配方
  硫酸   1 000 mL
  铬酐(99)   185~200 g·L^{-1}
  水   400 mL
  温度   55~60 ℃
  时间   30 min

③敏化液配方
  氯化亚锡(189.6)   10~15 g·L^{-1}
  盐酸   40 mL/L
  锡条(119)   1 根
  温度   室温
  时间   3~5 min

④活化液配方
  硝酸银(169.87)   1.5~2 g·L^{-1}
  氨水   滴至溶液透明
  温度   室温
  时间   3~5 min

⑤还原液配方
  甲醛（30）   (36%~38%)1 份
  水   9 份
  温度   室温
  时间   5~25 s

⑥化学镀铜液配方
  甲液:酒石酸钾钠(210.2)   40 g·L^{-1}
    a)氢氧化钠(40)   9 g·L^{-1}
    b)无水碳酸钠(106)   42 g·L^{-1}
  乙液:硫酸铜(160)   14 g·L^{-1}
    c)氯化镍(129.7)   4 g·L^{-1}
    d)甲醛(37%)   53 mL·L^{-1}
化学镀铜时按甲液:乙液＝3:1 混合使用。

⑦酸性光亮镀铜液配方
  硫酸铜   175~200 g·L^{-1}
  硫酸   60~70 g·L^{-1}
  氯离子   0.02~0.05 g·L^{-1}
  KG-5 光亮剂   5 mL·L^{-1}
  温度   10~40 ℃
  电流密度   4 A·dm^{-2}
  阳极   0.3%含磷铜板

实验19　过氧化氢分解催化剂的制备及其性能比较

【实验目的】

1. 制备 H_2-O_2 燃料电池的阴极催化剂,并通过其对 H_2O_2 的催化分解考察其催化活性;
2. 掌握量气法测定 H_2O_2 催化分解反应速率常数的动力学原理和方法。

【预习要求】

1. 理解 H_2O_2 分解反应与 H_2-O_2 燃料电池的关系;
2. 查阅文献,了解 H_2O_2 分解的催化剂;
3. 了解测定 H_2O_2 催化分解反应速率常数的动力学原理和方法。

【实验原理】

燃料电池是把化学能直接转化为电能的一种装置,它大大提高了能源转化率,并且可通过选择绿色燃料来避免污染源的产生。但燃料电池所发生的电化学反应由于受到反应速率的限制,大多很难用于实践。为克服这一困难,通常在装置中添加催化剂来控制反应速率,因此催化剂的选取对燃料电池的实际效用有着很重要的影响。

H_2O_2 是许多重要电化学反应(如 H_2-O_2 燃料电池中氧电极的电化学还原)的中间产物,其分解反应是电化学反应总反应的控制步骤。在 H_2-O_2 燃料电池中,氢负极交换电流可以很大,但氧正极交换电流较小,因为在室温下 O_2 在一般电极材料上还原很慢(H_2O_2 分解反应进行得很慢),因此必须选用有效的催化剂加速这一反应,才能显著地提高 H_2O_2 分解反应的速率,使燃料电池具有实用价值。

根据 O_2 电极反应机理,电催化反应最初进行单电子过程,生成 H_2O_2 中间产物,其反应为

$$\frac{1}{2}O_2 + H_2O + e^- \Longleftrightarrow \frac{1}{2}H_2O_2 + OH^-$$

生成的中间产物必须尽快分解,即

$$H_2O_2(aq) \xrightarrow{k} \frac{1}{2}O_2\uparrow + H_2O \qquad (4-19-1)$$

这样可降低其浓度以增加还原电势。因此,它是整个反应的控制步骤。

从催化动力学角度考虑,铂黑或银黑有很高的催化活性,但价格太高,而经研究发现,具有尖晶石结构的 $Cu_xFe_{3-x}O_4$、$Co_xFe_{3-x}O_4$ 等对 O_2 具有较高活性,且用常规方法制备这类催化剂并不困难。$Cu_{1.5}Fe_{1.5}O_4$ 作为这一类型催化剂,其制备工艺简单、操作方便,并且其制备原料容易获得,催化性能优异,应用在 H_2O_2 的催化分解实验中,既方便快捷,又效果明显。

制备 $Cu_{1.5}Fe_{1.5}O_4$ 的主要过程是先用 NaOH 溶液沉淀出 Cu(Ⅱ)和 Fe(Ⅲ)的氢氧化物,再将所得沉淀在空气中加热,进行氧化还原胶水,生成尖晶石结构的 $Cu_{1.5}Fe_{1.5}O_4$。

$$1.5CuCl_2 + 1.5FeCl_3 + 7.5NaOH = Cu_{1.5}Fe_{1.5}(OH)_{7.5} + 7.5NaCl$$

$$0.125O_2 + Cu_{1.5}Fe_{1.5}(OH)_{7.5} = Cu_{1.5}Fe_{1.5}O_4 + 3.75H_2O$$

催化剂的活性与沉淀速度、反应温度、胶水的温度和时间等有关。在碱性溶液中催化剂作用下 H_2O_2 按式(4-19-1)分解,因此可根据催化剂在碱性溶液中分解 H_2O_2 的能力来考察它的催化活性。式(4-19-1)已经证实为一级反应,其速率方程为

$$\ln \frac{C_t}{C_0} = -kt \quad (4-19-2)$$

式中,C_0 为 H_2O_2 的初始浓度;C_t 为 t 时刻 H_2O_2 的浓度;k 为反应的速率常数。

由式(4-19-1)可知,在 H_2O_2 的分解过程中,放出的氧气的体积与已分解的 H_2O_2 的浓度成正比。令 V_∞ 为 H_2O_2 全部分解生成的氧气的体积,V_t 为 H_2O_2 经分解时间 t 后生成的氧气的体积,f 表示一定时溶液中 H_2O_2 浓度与生成氧气的体积的比例常数,则有

$$V_\infty = fC_0 \quad V_\infty - V_t = fC_t \quad (4-19-3)$$

将式(4-19-1)带入式(4-19-2),整理后可得

$$\ln(V_\infty - V_t) = -kt + \ln V_\infty \quad (4-19-4)$$

若以 $\ln(V_\infty - V_t)$ 对 t 作图得一条直线,说明该反应是一级反应,由直线的斜率可求反应速率常数 k,通过控制实验温度,可求出不同温度下的反应速率常数 k,再根据阿仑尼乌斯公式

$$\ln \frac{k_2}{k_1} = \frac{E_a}{R}\left(\frac{T_2 - T_1}{T_1 T_2}\right)$$

可进一步求出反应的活化能 E_a,通过比较不同催化剂的表观活化能 E_a 的大小,来评价催化剂的活性。

V_∞ 可通过实验所用 H_2O_2 的浓度和体积算出。在酸性溶液中 H_2O_2 与 $KMnO_4$ 反应,即

$$5H_2O_2 + 2KMnO_4 + 3H_2SO_4 = 2MnSO_4 + K_2SO_4 + 8H_2O + 5O_2 \uparrow$$

用 $KMnO_4$ 标准溶液滴定加有一定量 H_2SO_4 的 H_2O_2 溶液,计算出 H_2O_2 溶液的浓度后,即可算出 H_2O_2 全部分解所生成的 O_2 的量,利用气体状态方程换算成实验条件下的 O_2 体积 V_∞(氧气分压为大气压减去实验温度下水的饱和蒸汽压)。

实验过程中,可改变催化剂的制备条件,如沉淀速度、反应温度、脱水的温度和时间等,比较不同实验条件下合成的催化剂的催化活性,也可将本实验合成的催化剂与其他 H_2O_2 分解催化的催化活性进行比较,进一步了解催化剂活性对 H_2O_2 分解速率的影响。

【仪器与试剂】

仪器:电子秒表,电磁搅拌器,烘箱,恒温水浴,循环水泵,布氏漏斗,真空干燥箱,量气管,250 mL 锥形瓶,滴定管,pH 试纸,移液管,三通旋塞,水位瓶。

药品:2% H_2O_2 溶液,1 mol·L^{-1} KOH 溶液,0.02 mol·L^{-1} $KMnO_4$ 标准溶液,3 mol·L^{-1} H_2SO_4 溶液,化学纯 $CuCl_2·6H_2O$,化学纯 $FeCl_2·6H_2O$,5 mol·L^{-1} NaOH 溶液,MnO_2 粉,CuO 粉,KI。

【设计要求】

1. 查阅文献,设计催化剂制备路线及动力学实验方法。
2. 计算温度为 T_1 时的 V_∞。
3. 分别以 $\ln(V_\infty - V_t)$ 为纵坐标,t 为横坐标作图,从所得直线的斜率求不同催化条件下 H_2O_2 分解反应的速率常数 k。
4. 在所用催化剂质量相同的条件下,根据其速率常数的大小,比较各催化剂的活性。
5. 改变实验温度至 T_2,重复上述实验,求出速率常数 k_2,并根据阿仑尼乌斯方程计算反应的表观活化能,并根据表观活性能的大小比较催化剂的催化活性。

【思考题】

1. 反应速率常数与哪些因素有关?
2. 还可用什么方法测定 V_∞?是否可消去 V_∞?
3. 你对本实验所用测定放出气体体积的方法有什么意见?

【相关阅读】

催 化 剂

催化剂又叫触媒。根据 IUPAC 于 1981 年提出的定义,催化剂是一种物质,它能够改变反应的速率而不改变该反应的标准 Gibbs 自由能变化。涉及催化剂的反应为催化反应。催化剂在化学反应中所起的作用叫催化作用。使化学反应加快的催化剂叫作正催化剂,使化学反应减慢的催化剂叫作负催化剂。

催化剂自身的组成、化学性质和质量在反应前后不发生变化,具有高度的选择性(或专一性),即一种催化剂并非对所有的化学反应都有催化作用,如二氧化锰在氯酸钾受热分解中可加快反应的速率,但对其他的化学反应就不一定有催化作用。氯酸钾热分解这个反应也并非只有二氧化锰这一种催化剂,氧化镁和氧化铜等也能催化氯酸钾快速分解。

催化剂在现代化学工业中占有极其重要的地位,据统计,约有 80%~85% 的化工生产过程使用催化剂(如氨、硫酸、硝酸的合成,乙烯、丙烯、苯乙烯等的聚合,石油、天然气、煤的综合利用,等等),目的是加快反应速率,提高生产效率。在科学实验和生命活动中,都离不开催化剂的作用。例如,合成氨中用铁为主的多分组催化剂,硫酸生产中用五氧化二钒作催化剂,提高反应速率。在炼油厂,催化剂更是少不了,选用不同的催化剂,就可以得到不同品质的汽油、煤油。汽车尾气中含有害的一氧化碳和一氧化氮,利用铂等金属作催化剂可以迅速将二者转化为无害的二氧化碳和氮气。酶是植物、动物和微生物产生的具有催化能力的蛋白质,生物体的化学反应几乎都在酶的催化作用下进行,酿造业、制药业等都要用催化剂催化。

根据催化反应进行是否有相界面存在,可以把催化剂分为均相催化剂和多相催化剂。

催化剂和反应物同处于一相,没有相界面存在的反应称为均相催化反应,能起均相催化作用的催化剂为均相催化剂。均相催化剂包括液体酸、碱催化剂,可溶性过渡金属化合物(盐类和络合物)等。均相催化剂以分子或离子独立起作用,活性中心均一,具有高活性和高选择性。

多相催化剂又称非均相催化剂,反应过程中催化剂和反应物、生成物处于不同相中,即和它们催化的反应物处于不同的状态。如在生产人造黄油时,通过固态镍(催化剂)能够把不饱和的植物油和氢气转变成饱和的脂肪。固态镍是一种多相催化剂,被它催化的反应物则是液态(植物油)和气态(氢气)。

目前,对催化剂的作用还没有完全弄清楚。在大多数情况下,人们认为催化剂本身和反应物一起参加了化学反应,降低了反应所需要的活化能。有些催化反应是由于形成了很容易分解的"中间产物",分解时催化剂恢复了原来的化学组成,原反应物就变成了生成物。有些催化反应是由于吸附作用,吸附作用仅能在催化剂表面最活泼的区域(叫作活性中心)进行。活性中心的区域越大或越多,催化剂的活性就越强。反应物里如有杂质,或吸附之后难以解吸,就可能使催化剂的活性减弱或失去,这种现象叫作催化剂的中毒。

实验20　植物色素热降解动力学参数的测定

【实验目的】

1. 熟悉化学反应速率方程的测定方法及思路;
2. 掌握速率常数、半衰期及活化能的求解;
3. 掌握分光光度计的使用方法;
4. 通过本实验,加深对动力学知识的理解,并学会灵活运用化学动力学手段解决实际问题。

【预习要求】

1. 查阅文献,设计出植物色素的提取方案及详细步骤;
2. 查阅相关资料,确定色素降解过程中浓度的测定方法;
3. 了解速率常数、半衰期及活化能的概念及求解方法。

【实验原理】

色、香、味、形是构成食品感官的四大要素,良好的色泽使人赏心悦目,刺激人的食欲,但食品加工过程往往出现不可避免的色泽变化或褪色现象,影响产品的感官品质。为此,食品加工中需要人为地添加食品着色剂(食用色素)以恢复或改善其本来的色泽。天然色素以其天然、绿色、安全等优点被大多数消费者所接受和喜欢。然而,天然色素普遍存在稳定性差的问题,在自然存放的过程中,易受光、热、氧、pH值及共存物等的影响而褪色或变色,在一定程度上限制了天然色素的利用。对天然色素进行热、光、氧降解的动力学研究可以了解色素的稳定性,为天然色素的应用奠定基础。下面以龙葵色素为例进行说明。

龙葵为茄科茄属一年生草本植物,在东北地区分布广泛,据资料介绍,长白山区野生龙葵约有 5×10^4 t,有很大的开发潜力。龙葵红色素属花色苷类天然色素,提取工艺简便,色泽鲜艳,品质优良,无毒副作用,着色自然且营养丰富。花色苷被证明具有医药功效,包括抗氧化(抑制脂质过氧化、清除自由基活性、体内抗氧化)作用,抗炎和抗昏厥活性,降低血清脂质水平,抗变异与抗肿瘤作用,改善肝功能效果等,是很有开发潜力的天然食品色素品种,但该色素对热、光、氧等敏感。

龙葵色素的稳定性是开启其应用市场大门的钥匙,故正确认识其不稳定因素,提高其稳定性和应用价值具有重要意义。龙葵红色素的热降解动力学参数的测定属于热稳定性研究范畴,是确定其提取和浓缩条件及食品加工中应用的基础。

本实验是依据经典的物理学和物理化学理论——朗伯-比尔(Lambert-Beer)定律、一级化学反应动力学方程和 Arrhenius 方程开展色素热降解程度测定和表征热降解规律的。

龙葵红色素降解反应符合一级反应动力学。在一定波长的单色光照射下,处于酸性的龙葵色素液的浓度与吸光度 A 成正比(该色素提取液在波长 514~530 nm 有最大吸收),在色素热降解过程中,随着色素浓度的减小,吸光度值不断减小,而色素浓度随 t 的变化符合一级反应动力学规律。因此,可用色素液吸光度代替其浓度,通过测定吸光度值的方法确定其降解程度,得出反应速率方程。实验测定数据并做 $\ln A - t$ 曲线,得到色素热降解反应速率方程,进而得到速率常数 k 和半衰期 $t_{1/2}$,再作 $\ln k - 1/T$ 曲线,又可由其斜率和 Arhenius 方程求得色素热降解反应活化能。

【仪器与试剂】

仪器:分析天平1台,分光光度计1台,精密酸度计2台,离心机1台,电冰箱,磁力加热搅拌器1台,恒温水浴1台,称重计1台,必要的玻璃仪器。

药品:鲜龙葵果,去离子水,柠檬酸(分析纯),无水乙醇(分析纯)。

【设计要求】

1. 查阅文献,根据本实验所提供的仪器设计出提取龙葵色素的实验方案并进行实验。
2. 采用分光光度计测试龙葵红色素的最大吸收波长。
3. 测定不同温度时最大吸收波长下的吸光度值,得出热降解反应速率方程。
4. 求出该色素热降解反应的反应速率常数、半衰期和活化能等动力学参数。

【数据记录与处理】

1. 本实验测定 40 ℃ 与 60 ℃ 两个温度下龙葵红色素的稳定性,所得实验数据记录于表 4-20-1 和表 4-20-2 中。

表 4-20-1 40 ℃时色素液稳定性实验数据

时间/h	吸光度值 A_1	吸光度值 A_2	吸光度值 A_3	平均吸光度值 A	ln$A$
0					
1					
2					
3					
4					
5					
7					
9					
11					
13					
15					
17					
20					
⋮					

表 4-20-2 60 ℃时色素液稳定性实验数据

时间/h	吸光度值 A_1	吸光度值 A_2	吸光度值 A_3	平均吸光度值 A	ln$A$
0					
1					
2					
3					
4					
5					
7					
9					
11					
13					
15					
17					
20					
⋮					

2. 根据表 4-20-1 和表 4-20-2 中的数据，绘图计算得出不同温度下的速率常数及半衰期，并根据不同温度下的速率常数绘图或计算得出龙葵红色素热降解的活化能。

【思考题】

1. 影响吸光度数据准确性的因素有哪些？

2. 为何 pH 值对龙葵红色素有影响？
3. 怎样通过实验数据来判断龙葵红色素热降解反应是否为一级反应？
4. 测定龙葵红色素热降解反应活化能时采用的温度区间是越大好还是越小好，为什么？
5. 许多天然色素都有热降解性。可根据季节选择合适的植物花、叶及果提取色素，如红龙果、红苋菜叶、落葵果、葡萄皮、山竹皮、红番茄、紫茄皮、黑米等。

【相关阅读】

1. 化学反应动力学参数的确定方法

动力学速率方程一般可归纳为

$$-\frac{dC_A}{dt} = kC_A^n \tag{4-20-1}$$

在这类方程中，动力学参数只有 k 和 n，故所谓速率方程的确定就是确定这两个参数，但方程的形式只取决于 n，k 不过是式中的一个常数，所以确定速率方程的关键是确定反应级数。

为了确定反应级数，需要有一定温度下不同时刻 t 的反应物浓度 C_A 的数据，即需要知道化学反应的 $C_A - t$ 关系。有了这一关系，就可以求得 n，进而求得 k。因此，要获得化学反应动力学参数即要准确测定反应进行到不同时刻反应物或生成物的浓度，而这一点常常是很困难的，尤其是对那些反应速率很快的反应。通常测定物质的浓度有两种方法，即化学分析法和物理化学分析法。

（1）化学分析法：反应进行后，在某一时刻突然使反应终止（如用骤冷反应系统，冲稀或迅速除去反应物质等），再取样，做化学分析，测定反应系统中物质的浓度。化学分析法的优点是能直接测出反应组分的浓度，缺点是手段烦琐，另外中断反应对连续生产不利。

（2）物理化学分析法：如果反应物和生成物的某种物理性质有较大的差别，则随着反应进行，这种物理性质就不断出现较显著的改变。在不中断反应进行过程的情况下，测量不同时刻反应系统中这一性质的改变情况，从而间接取得浓度变化的情况，就可计算出浓度的变化和化学反应速率。通常利用的物理性质有压力、体积、旋光度、比色、吸光度、折射率、电导、电动势和光谱等。物理化学分析的方法可以连续进行，不像化学方法那样要中断反应，但须先确定浓度-物理性质间的关系，即做一系列的标准曲线。

获得了反应物浓度随时间变化关系后，即可根据微分法、尝试法或半衰期法等来求得 k 和 n 等动力学参数。

（1）微分法 对式（4-20-1）取对数得

$$\lg\left(-\frac{dC_A}{dt}\right) = n\lg C_A + \lg k \tag{4-20-2}$$

以 $\lg\left(-\frac{dC_A}{dt}\right)$ 对 $\lg C_A$ 作图为一直线，直线的斜率 $m = n$，截距 $b = \lg k$。

微分法的几种特殊处理方法如下。

① 初始速率法。有时反应产物对反应速率会有影响，为消除产物的影响故采用此法。在一定速率下，反应开始时的瞬时速率为初始速率。由反应物浓度的变化确定反应速率和速率方程式的方法为初始速率法。具体操作：将反应物按不同组成配制成一系列混合物，

先只改变一种反应物 A 的浓度,保持其他反应物浓度不改变。在某一温度下反应开始,获得 $C_A - t$ 图,确定 $\Delta t \to 0$ 时的瞬时速率。若能获得至少两个不同 C_A 条件下的瞬时速率,即可确定反应物 A 的反应级数。同样的方法可以确定其他反应物的反应级数。

②隔离法。若一个反应速率方程可写为

$$-\frac{\mathrm{d}C_A}{\mathrm{d}t} = k' C_A^\alpha C_B^\beta \qquad (4-20-3)$$

让除 A 以外的其他组分浓度大大过剩,则速率方程近似化简为

$$-\frac{\mathrm{d}C_A}{\mathrm{d}t} = k C_A^\alpha \qquad (4-20-4)$$

这时求出的级数 α 是组分 A 的级数,而不是反应的总级数。

③按反应计量系数比投料法。若一个反应速率方程可写为

$$-\frac{\mathrm{d}C_A}{\mathrm{d}t} = k' C_A^\alpha C_B^\beta \qquad (4-20-5)$$

按反应计量系数比投料,反应速率可简化为

$$-\frac{\mathrm{d}C_A}{\mathrm{d}t} = k C_A^{\alpha+\beta+\cdots} = k C_A^n \qquad (4-20-6)$$

这样所求级数为反应总级数。

(2) 尝试法(试差法)。尝试法包括代入尝试法和作图尝试法,一般适用于整数级。

①代入尝试法。把实验测得的多组 C_A、t 数据分别代入简单级数反应的积分方程关系中,看按哪一级数的积分式计算出来的 k 为常数,即为哪级反应。

②作图尝试法。把实验测得的 C_A、t 数据按简单级数反应的线性关系方程作图,符合哪个级数直线关系即为哪级反应。

此外还有半衰期法,可用半衰期法求除一级反应以外的其他反应的级数。

2. 花色苷热降解动力学介绍

花色苷是 18 种天然存在的花色素的糖苷化合物,为佯盐离子的多羟基及多甲基衍生物。花色苷在有氧(甚至无氧)、pH 值变化及热、光、酶等作用下易降解。降解过程为佯盐离子→假碱型→查尔酮→醌碱型。龙葵红色素为花色苷类色素,花色苷在低温下的稳定性较好,加热会使花色苷向着无色的查尔酮结构移动,在 pH 值下,颜色由鲜艳的紫红色逐渐变暗,特定波长下吸光度值下降,此现象反映其发生热降解。该色素长时间加热会发生显著热降解,但食品加工时,若在 100 ℃下加热时间少于 12 min,花色苷的损失可忽略不计。在采取措施排除光照和氧化干扰后,假定龙葵红色素热降解:①无光干扰;②无空气氧化干扰;③无酶(糖苷酶和多酚氧化酶)干扰;④实验条件控制精度满足稳定地进行热降解反应需要。

文献报道花色苷热降解为一级反应,因在特定波长下,花色苷浓度与吸光度值成正比,故用吸光度值 A 代替色素浓度的一级反应速率方程积分式应为 $\ln A = -kt + B$(B 为与反应体系和条件有关的常数),色素热降解反应的半衰期应为 $t_{1/2} = 0.693/k$。在某温度下,由测定的 $A - t$ 数据对 $\ln A - t$ 作图并进行线性回归,即可得到该温度下的 $\ln A - t$ 回归方程以及 k 和 $t_{1/2}$。

由 Arrhenius 公式 $\ln k = \frac{E_a}{R} \cdot \frac{1}{T} + \ln A$ 作图,或用公式 $\ln \frac{k_2}{k_1} = -\frac{E_a}{R} \left(\frac{1}{T_2} - \frac{1}{T_1} \right)$ 计算求得色素热降解反应的活化能 E_a,进一步可由 E_a 数值的大小粗略判断提取色素的稳定性。若 E_a

很小,在实验条件下色素极不稳定;若 E_a 较小,在实验条件下色素较不稳定;若 E_a 数值不大不小,在实验条件下色素可稳定存在;若 E_a 数值较大,在实验条件下色素很稳定。一般龙葵红色素降解反应活化能并不太小,热稳定性不是很差,若控制好 pH 值,在缺氧、避光、灭酶前提下提取并负压浓缩,一般温度下还是比较稳定的。

虽然龙葵红色素受热降解是一个复杂过程,其降解机理目前尚不十分清楚,但用热降解化学动力学规律是完全可以间接反映其热稳定性的。本实验的结果对龙葵红色提取液的后加工及应用有重要参考价值。

Experiment 20 Determination of Kinetic Parameters of Thermal Degradation of Plant Pigments

【Purpose of the Experiment】

1. Familiar with the determination method and idea of chemical reaction rate equation;
2. Master the solution of rate constant, half-life and activation energy;
3. Master the use of spectrophotometer;
4. Through this experiment, understanding the kinetics knowledge and learning to use chemical kinetics flexibly to solve practical problems problem.

【Preview Requirements】

1. Consult the literature and design the extraction plan and detailed steps of plant pigments;
2. Consult relevant information to determine the method for determining the concentration of the pigment in the degradation process;
3. Understand the concepts and solution methods of rate constant, half-life and activation energy.

【Principle】

The color of food is closely related to the processing and credit performance of the food. Brightly colored food can increase appetite and enhance the enjoyment of eating.

Color, aroma, taste and shape are the four major elements that make up the senses of food. Good color makes people please to the eye and stimulates people's appetite. However, the inevitable color change or fading phenomenon often occurs during food processing, which affects the sensory quality of the product. For this reason, food colorants (food pigments) need to be added artificially in food processing to restore or improve its original color. Natural pigments are accepted and liked by most consumers for their natural, green, safe and other advantages. However, natural pigments generally have the problem of poor stability. In the process of natural storage, they are susceptible to fading or discoloration under the influence of light, heat, oxygen, pH and coexisting substances, which limits the use of natural pigments to a certain extent. The kinetic study of the degradation of natural pigments by heat, light and oxygen can understand the stability of pigments and lay the foundation for the application of natural pigments. Take Solanum pigment as an example for illustration.

Solanum is an annual herb of Solanaceae. It is widely distributed in Northeast China. According to the data, the wild Solanum in Changbai Mountains is about 5×10^4 t, which has great development potential. Solanum red pigment is a natural anthocyanin pigment. It has simple extraction process, bright color, good quality, no toxic side effects, natural coloration and rich nutrition. Anthocyanins are proven to have medicinal effects, including anti-oxidation (inhibition of lipid peroxidation, scavenging free radical activity, anti-oxidation in the body), anti-inflammatory and anti-fainting activities, lowering serum lipid levels, anti-mutation and anti-tumor effects, and improving liver function effects, etc, which are natural food pigments with great development potential, but the pigments are sensitive to heat, light, and oxygen. The stability of solanum pigment is the key to open the door to its application market, so it is of great significance to correctly understand its unstable factors and improve its stability and application value. The determination of thermal degradation kinetic parameters of solanum red pigment belongs to the category of thermal stability research and is the basis for determining its extraction and concentration conditions of food processing applications.

This experiment is based on the classic physics and physical chemistry theory-Lambert Beer law, first-order chemical reaction kinetic equation and Arrhenius equation to determine the degree of thermal degradation of pigments and characterize the thermal degradation laws.

The degradation reaction of solanum red pigment accords with the first-order reaction kinetics. Under the irradiation of a certain wavelength of monochromatic light, the concentration of the acidic solanum pigment solution is proportional to the absorbance A (the pigment extract has a maximum absorption between the wavelength of 514 ~530 nm). During the thermal degradation of the pigment, as the concentration of the pigment decreases, the absorbance value continues to decrease, and the change of pigment concentration with t conforms to the first-order reaction kinetics. Therefore, the absorbance of the pigment liquid can be used instead of its concentration, and the degree of degradation can be determined by measuring the absorbance value, and the reaction rate equation can be obtained. The experimental data is measured and the $\ln A \sim t$ curve is made to obtain the thermal degradation reaction rate equation of the pigment, and then the rate constant k and the half-life $t_{1/2}$ are obtained, and then the $\ln k \sim 1/t$ curve can be obtained. Finally, the activation energy of the thermal degradation reaction of the pigment can be obtained from its slope and Arrhenius equation.

【Apparatus and Reagents】

Instruments: One analytical balance, one spectrophotometer, two precision acidity meters, one centrifuge, one refrigerator, one magnetic heating stirrer, one constant temperature water bath, one weighing meter, and necessary glass instruments.

Reagents: Fresh solanum fruit, deionized water, citric acid (analytical grade), absolute ethanol (analytical grade).

【Design requirements】

1. Consult the literature, design an experiment plan for extracting Solanum pigment and carry out the experiment according to the equipment provided in this experiment.

2. Use a spectrophotometer to test the maximum absorption wavelength of the solanum red pigment.

3. Determine the absorbance value under the maximum absorption wavelength at different temperatures, and process and regress the data to obtain the thermal degradation reaction rate equation.

4. Calculate the kinetic parameters such as the reaction rate constant, half-life and activation energy of the thermal degradation reaction of the pigment.

【Data Recording and Processing】

1. In this experiment, the stability of the solanum red pigment under two temperatures of 40 ℃ and 60 ℃ should be determined, and the experimental data obtained are recorded in the Table 4 – 20 – 1 and Table 4 – 20 – 2.

Table 4 – 20 – 1 Experimental data of liquid pigment stability at 40 ℃

Time/h	Absorbance value A_1	Absorbance value A_2	Absorbance value A_3	Average absorbance value A	ln A
0					
1					
2					
3					
4					
5					
7					
9					
11					
13					
15					
17					
20					
⋮					

表 4 – 20 – 2 Experimental data of liquid pigment stability at 60 ℃

Time/h	Absorbance value A_1	Absorbance value A_2	Absorbance value A_3	Average absorbance value A	ln A
0					
1					
2					
3					
4					
5					
7					
9					
11					
13					
15					
17					
20					
⋮					

2. According to the data in Table 4 – 20 – 1 and Table 4 – 20 – 2, the rate constants and half-lives at different temperatures can be calculated by plotting, and the activation energy of degradation for solanum red pigment can be calculated by plotting the rate constants at different temperatures.

【Questions】

1. What are the factors that affect the accuracy of absorbance data?

2. Why does pH affect the solanum red pigment?

3. How to judge whether the thermal degradation reaction of solanum red pigment is a first-order reaction through experimental data?

4. Whether the temperature range used to determine the activation energy of the thermal degradation reaction of solanum red pigment is larger or smaller, and why?

5. Many natural pigments are thermally degradable. According to the season, choose suitable plant flowers, leaves and fruits to extract pigments, such as red dragon fruit, red amaranth leaves, basil fruit, grape skin, mangosteen skin, red tomato, purple eggplant skin, black rice, etc.

Experiment 21 Nano-TiO$_2$ Prepared by Sol-gel Method and Its Photocatalytic Activity

【Objectives】

1. Understanding the prepared method and experiment condition of nano-TiO$_2$.
2. Understanding the photocatalytic principle of nano-TiO$_2$.
3. Understanding the UV-vis spectrophotometer and data processing method.

【Principles】

Organic pollution has for decades become an increasingly serious problem in environmental field. Among various methods of organic pollutants treating, photocatalysis technology has attracted enormous attention because of its high efficiency, energy-saving, environmental friendly property and so on. As one of the excellent candidates in semiconductor photocatalysis, TiO$_2$ has been investigated extensively due to its low cost, easy availability, long-term chemical stability and environmental benignity. The preparation methods of nano-TiO$_2$ can be divided into physical method and chemical method. Compared with simple physical method, chemical method is an important method to prepare nano-TiO$_2$. Among many chemical methods, sol-gel method has the advantages of good chemical uniformity, high purity, easy operation and low cost, etc. Therefore, this experiment adopts sol-gel method to prepare nano-TiO$_2$.

In the sol-gel process, the butyl titanate [Ti(OC$_4$H$_9$)$_4$] is used as precursor, anhydrous ethanol (C$_2$H$_5$OH) is used as solvent and acetic acid (CH$_3$COOH) is used as chelating agent, and the detailed reaction are as follows:

$$\text{Ti(OC}_4\text{H}_9)_4 + 4\text{H}_2\text{O} \rightarrow \text{Ti(OH)}_4 + 4\text{C}_4\text{H}_9\text{OH}$$
$$\text{Ti(OH)}_4 + \text{Ti(OC}_4\text{H}_9)_4 \rightarrow 2\text{TiO}_2 + 4\text{C}_4\text{H}_9\text{OH}$$
$$\text{Ti(OH)}_4 + \text{Ti(OH)}_4 \rightarrow 2\text{TiO}_2 + 4\text{H}_2\text{O}$$

During the photocatalytic process, when nano-TiO$_2$ is excited by the UV light, the photogenerated electrons (e$^-$) are excited from the valence band of TiO$_2$ to the conduction band, and the photogenerated holes (h$^+$) are left in valence band (Figure 4-21-1). The produced electrons (e$^-$) and holes (h$^+$) can migrate to the surface of TiO$_2$ and take part in the following reactions:

$$\text{h}^+ + \text{e}^- \rightarrow \text{recombination}$$
$$\text{h}^+ + \text{organic pollutants} \rightarrow \text{CO}_2 + \text{H}_2\text{O}$$
$$\text{H}_2\text{O} + \text{h}^+ \rightarrow \cdot\text{OH} + \text{H}^+$$
$$\text{O}_2 + \text{e}^- \rightarrow \cdot\text{O}_2^+$$
$$\cdot\text{OOH} + \text{H}_2\text{O} + \text{e}^- \rightarrow \text{H}_2\text{O}_2 + \text{OH}^-$$

$$H_2O_2 + e^- \rightarrow \cdot OH + OH^-$$
$$H_2O_2 + \cdot O_2^+ \rightarrow \cdot OH + OH^- + O_2$$
$$\cdot OH + \text{organic pollutants} \rightarrow CO_2 + H_2O$$

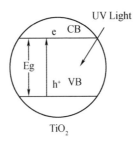

Figure 4-21-1 The energy band diagram of TiO_2

In this experiment, nano-TiO_2 powders are prepared by sol-gel method firstly, and its photocatalytic activity is evaluated by photodegradation rate of methyl orange (MO) solution.

【Apparatus and reagents】

Analytical balance; Centrifugal machine; Centrifuge tube; Magnetic stirrers; Oven; 300 W mercury lamp; Ultraviolet-visible spectrophotometer; Measuring cylinder; Beaker; Magneton.

【Design Requirements】

1. Check the literature and design an experiment plan for "Nano-TiO2 prepared by sol-gel method and its photocatalytic activity" based on the equipment provided in this experiment. The designed contents should contain the synthesis of TiO2 powder, the tests of photocatalytic activites and recording and analyzing data.

2. Carry out experiments according to the designed experimental plan.

3. Process experimental data. Analyze experimental results.

4. Evaluate the catalytic activity according to results of photocatalytic activites.

【Post Lab Questions】

1. What are the roles of CH_3COOH and C_2H_5OH in the preparation process of nano-TiO_2?

2. What are the evaluation systems for photocatalysts based on the reported literatures?

Experiment 22 Determination of Pd/C Catalytic Performance for Hydrogen Evolution Reaction (HER)

【Purpose of the Experiment】

1. To know the basic principles of HER.
2. To learn the test method of HER.

【Principle】

Hydrogen, as a clean and potentially renewable energy source, has been considered as a promising alternative to replace fossil fuel if hydrogen can be produced cleanly by electrolysis or photolysis of water. A low production efficiency and high cost limit the use of hydrogen energy. However, water electrolysis has got a promising hydrogen production method because it is suitable for large-scale production, and has high production efficiency and high purity of products, already attracting wide attention. The hydrogen evolution reaction (HER) for water splitting contains two pathways: the Volmer-Heyrovsky pathway and the Volmer-Tafel pathway.

In acidic solution:

$$H_3O^+ + e^- + * \rightarrow H^* + H_2O \quad (4-22-1)$$
$$H_3O^+ + e^- + H^* \rightarrow H_2 + H_2O \quad (4-22-2)$$
$$H^* + H^* \rightarrow H_2 \quad (4-22-3)$$
$$H^+ + H^+ + 2e^- \rightarrow H_2 \quad (4-22-4)$$

In alkaline solution:

$$H_2O + e^- \rightarrow H^* + OH^-$$
$$H_2O + e^- + H^* \rightarrow H_2 + OH^-$$
$$H^* + H^* \rightarrow H_2$$

Currently, precious metals are considered is to be one of the most excellent electrocatalysts for HER. The catalytic activity and stability of the catalyst are the critical issues. The polarization curve and the Tafel slope are generally used to evaluate the activity of the catalyst. Chronoamperometry is carried out to characterize the stability of the catalyst.

When testing the HER of commercial Pd/C, the acidic or basic electrolyte solutions can be selected. However, the corresponding reference electrodes are different. In the H_2SO_4 solution, Ag/AgCl electrode or saturated calomel electrode (SCE) is often used as the reference electrode. When SCE used, the measured voltage was calibrated to a reversible hydrogen electrode (RHE) according to E vs. RHE (V) = E vs. SCE (V) + 0.245 + 0.059 1 × pH. When Ag/AgCl electrode electrodes were used, the conversion was performed by E vs. RHE(V) = E vs. Ag/AgCl (V) + 0.198 9 + 0.059 1 × pH. The Hg/HgO electrode is selected as the reference electrode in a alkaline solution and converted to the standard hydrogen potential by E vs. HER (V) = E vs.

Hg/HgO (V) +0.098 +0.0591×pH.

【Apparatus and Reagents】

CHI Electrochemical workstation 1
Reference electrode, SCE 1
Auxiliary electrode, Pt wire electrode 1
H_2SO_4 solution (0.5 mol/L)
KOH solution (1.0 mol/L)
Glass carbon electrode (GCE) 1

【Design Requirements】

1. Check the literature and design an experiment plan for "Determination of Pd/C Catalytic Performance for Hydrogen Evolution Reaction" based on the equipment provided in this experiment. The designed contents should be contain the synthesis of Pd/C catalysts, the pretreatment of glass carbon electrode, the preparation of the working electrode, the tests of Linear Scan Voltammetry and Chronoamperometry, processing and analyzing data.

2. Carry out experiments according to the designed experimental plan.

3. Process experimental data. Analyze experimental results.

4. Evaluate the catalytic activity according to Linear Scan Voltammetry and the stability according to Chronoamperometry.

【Post Lab Questions】

1. What do you think about the overpotential at the current density of 10 mA/cm^2 and the stability for Pd/C catalyst in a alkaline solution?

2. Is the HER performance affected by the electrochemical reaction temperature?

第5章 仪 器

仪器1 恒温水浴

恒温水浴的结构及工作原理如图5-1-1所示。

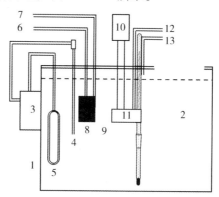

1—槽体;2—恒温介质;3—控制器;4—温度传感器;5—加热器;6—冷却水入口;7—冷却水出口;
8—蛇形制冷器;9—温度计;10—电机;11—搅拌器兼水泵;12—恒温水出口;13—恒温水返回口。

图5-1-1 恒温水浴的结构及工作原理图

槽体用于储存恒温介质。如果控制温度与室温相差不大,可用敞口大玻缸作为浴槽,对于较高或较低温度,应考虑保温问题。具有循环泵的恒温水浴,有时仅作存储、供给恒温液体之用,而实验在另一工作槽内进行。这种利用液体循环恒温的工作槽可做得小一些,以减小温度控制的滞后性。

搅拌器用于搅拌恒温介质,使恒温水浴温度均匀。搅拌器的功率安装位置和桨叶的形状对搅拌效果有很大影响,搅拌器应装在加热器上面或靠近加热器,使加热后的液体及时混合均匀再流至恒温区,可以用循环水流代替搅拌。

加热器用于向槽中供给热量以补偿散失的热量,通常采用电加热器间歇加热,要求热容量小、导热性好、功率适当。

制冷器用于从恒温水浴取走热量,以抵偿槽中增加的热量。

温度传感器用于探测恒温介质的温度,控制加热器和制冷器是否工作,如接触温度计(或称水银定温计)、热敏电阻感温元件等。

恒温控制器起一个继电器的作用。恒温介质的温度高于设定温度时,继电器切断加热电路,接通制冷电源;恒温介质的温度低于设定温度时,继电器接通加热电路,切断制冷电源。

1. 恒温水浴的灵敏度

恒温水浴的温度控制装置属于"通""断"类型,当加热器接通后,恒温介质温度上升,热量传递需要时间,所以会出现温度传递的滞后,而加热器附近的温度超过指定温度的现象。同理降温时也会出现滞后现象。恒温水浴控制的温度有一个波动范围,波动范围越小,恒温水浴的灵敏度越高。灵敏度与感温元件、继电器性能、搅拌器的效率、加热器的功率等因素有关。

恒温水浴灵敏度的测定是在指定温度下用较灵敏的温度计(贝克曼温度计或精密温差仪)记录温度随时间的变化,以温度为纵坐标、时间为横坐标绘制成温度-时间曲线,如图5-1-2所示。

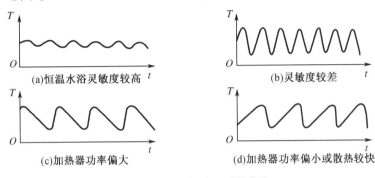

图 5-1-2 温度-时间曲线

恒温水浴灵敏度 t_E 与最高温度 t_1、最低温度 t_2 的关系式为

$$t_E = \pm \frac{t_1 - t_2}{2}$$

2. CS501 型恒温水浴

CS501 型恒温水浴是实验室最常用的恒温水浴,用于精密恒温,如图5-1-3所示。

恒温水浴上装有两组发热元件,分别为 500 W 和 1 000 W,发热快,余热少。特别应注意在槽内未加入恒温介质前,切勿通电,以防因温度过高,将加热护管烧坏。

冷凝管 1 个,由此通入冷却水,有进水嘴、出水嘴各 1 只,固定于恒温水浴盖板上。

电动水泵 1 个,在控制箱上有单独开关,2 800 r·min^{-1},电动机与水泵用螺旋形钢丝相连,抽水量 4 L/min 左右,可以将恒温水输送至恒温水浴外需要恒温的仪器设备。水泵转轴尾部有搅拌叶子一只,作液体循环之用,使恒温水浴内温度均匀。

接触温度计 1 支,是温度传感器。如图5-1-4所示,其结构与普通水银温度计近似,在毛细管上部悬有一根可上下移动的金属丝,从水银槽也引出一根金属丝,两根金属丝再与温度控制系统连接。在其上部装有一根可随管外永久磁铁帽旋转的螺杆,螺杆上有一指示金属片(标铁),标铁与毛细管中的金属丝(触针)相连,当螺杆转动时标铁上下移动即带动金属丝上升或下降。调节温度时,先转动调节帽,使螺杆转动,带着金属片移动至所需温度的刻度位置。当加热器加热,水银柱上升与金属丝相接,两根导线连通,使继电器动作,加热器电源被切断,停止加热。当恒温水浴温度下降,水银柱下降与金属丝断开,两根导线不通,使继电器动作,加热器电源被接通,开始加热。由于接触温度计的温度刻度很粗糙,恒温水浴的精确温度应该由另一精密温度计指示。当所需的控温温度稳定时,将调节帽上的固定旋钮旋紧,使之不发生转动。接触温度计允许通过的电流很小,约为几毫安以下,不能同加热器直接相连,必须通过继电器与加热器相连。

继电器安装于控制器箱内,根据接触温度计的信号,控制加热元件是否工作。

CS501 型恒温水浴技术规格:电源 220 V,50 Hz;使用温度范围 < 95 ℃;温度波动 ±0.05 ℃;电动机功率 40 W;水泵流量 4 L·min^{-1}。

CS 501 型恒温水浴使用方法如下:

(1)在恒温水浴内灌注蒸馏水(或去离子水),水面至盖板 3 ~ 5 cm,不能过高,防止溢出;不能过低,防止缺水烧坏电热管。特别注意恒温水浴内不能使用自来水,否则会在筒壁和电热管上积聚水垢而影响恒温灵敏度。

(2)检查水泵出水口和进水口是否连接好。如果恒温水浴外的仪器设备需要恒温,将恒温水浴水泵出水口与待恒温仪器进水口连接,恒温水浴水泵进水口与待恒温仪器出水口连接。如果不需将恒温水输送至恒温水浴外,必须将恒温水浴水泵出水口与进水口连接,否则恒温水会喷出恒温水浴外。

(3)调节接触温度计。将接触温度计上端的马蹄形磁铁调节帽缓缓左右旋动,使温度计内调节螺杆旋转,先将标铁调到比所希望控制的温度低 1 ~ 2 ℃。

(4)使用恒温水浴需接地线。打开电源开关,打开水泵开关,使槽内水循环对流;打开加热器开关,加热指示灯亮表示处于加热状态(恒温水浴内温度低于设定温度);加热指示灯灭表示处于停止加热状态(恒温水浴内温度高于设定温度),加热器停止工作。

(5)当加热指示灯熄灭时(或一会儿亮一会儿灭),打开冷却水开关。

(6)观察恒温水浴上的另一精密温度计,如果其指示的温度低于设定温度,调节接触温度计上的标铁上移;如果其指示的温度高于设定温度,调节接触温度计上的标铁下移。反复进行调节,直到精密温度计指示的温度达到所需设定的温度,同时加热指示灯处于一会儿亮一会儿灭的状态。此时将接触温度计调节帽上的锁紧螺丝拧紧。

(7)调好后在使用过程中不能关掉搅拌器开关及冷却水。

(8)使用完毕,关闭各开关,关冷却水。

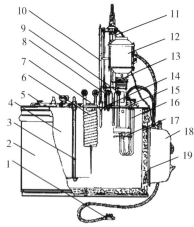

图 5 - 1 - 4 接触温度计

1—电源插头;2—槽体外壳;3—活动支架;4—恒温桶;5—恒温桶加水口;
6—冷凝管及进出水口;7—盖子;8—水泵进水;9—水泵出水;10—温度计;
11—接触温度计;12—电机;13—水泵;14—加水口;15—加热元件接线柱;
16—加热元件;17—搅拌叶;18—控制器;19—保温层。

图 5 - 1 - 3 CS501 型恒温水浴示意图

3. HK-1D 型恒温水浴

HK-1D 型恒温水浴如图 5-1-5 所示,集智能化控温器、玻璃恒温水浴、电动搅拌机构于一体,具有控温精度高、体积小、使用方便等优点。

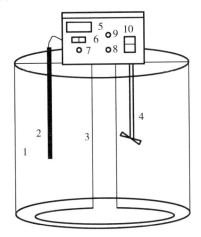

1—玻璃缸水浴;2—温度传感器;3—电热管;4—搅拌器;5—显示框;
6—测量/设定选择;7—温度设定;8—调速旋钮;9—加热指示灯;10—电源开关。

图 5-1-5 HK-1D 型恒温水浴

HK-1D 型恒温水浴的组成及性能:圆形玻璃缸,直径 300 mm,深 300 mm;加热器,电源 220 V,50 Hz;加热功率 1 kW;电动搅拌机,功率 35 W,无级调速,转速 0~400 /min 可调;智能化控温单元,电源 220 V,50 Hz,可控功率 0~1 kW,控温范围为 20~99 ℃,控温精度 ±0.05℃,控温稳定度 ±0.01 ℃。

HK-1D 型恒温水浴使用方法如下:

(1)在玻璃缸中加入去离子水至加热器上方,约至玻璃缸 4/5 深度,水位不能过低,以防烧坏加热管。

(2)恒温水浴必须接地。先将"测量/设定"开关置于设定,搅拌器的"调速"旋钮逆时针调到底(转速为零),然后打开电源开关。

(3)通过"调速"旋钮调节合适的搅拌速度。

(4)将"测量/设定"开关置于设定位置,通过"温度设定"旋钮设定温度值,然后将"测量/设定"开关置于测量位置,控制系统将自动加热水浴并控制在设定温度。

(5)控温仪的控温精度为 ±0.05 ℃,分辨率为 0.01 ℃。当加温到设定的温度附近时,控温仪上的加热指示灯不停地闪动。

仪器 2 阿贝折射仪

当光线从一种介质 B 进入另一种介质 A 时,由于两种介质的光学性质不同,在界面上会发生光的折射,如图 5-2-1 所示。对任何两种介质,在一定波长和温度下,入射角 i 的

正弦与折射角 γ 的正弦之比等于它在两种介质中传播速率 v_A、v_B 之比,即

$$\frac{\sin i}{\sin \gamma} = \frac{v_B}{v_A} = n_{B,A}$$

$n_{B,A}$ 称为折射率。在一定波长和温度条件下,折射率是物质的特性常数,故对指定的两种介质,折射率 $n_{B,A}$ 为一定值。折射率与波长、温度有关,在其右下角注以字母,表示测定时所用单色光的波长,D,F,G,C…分别表示钠的 D (黄)线,氢的 F(蓝)线、G(紫)线、C(红)线等;在其右上角注以测定时的介质温度(℃),如 n_D^{20} 表示 20℃时介质对钠光 D 线的折射率。

光线由 B 进入 A 时,入射角 i 大于折射角 γ,$n_{B,A} > 1$。当入射角 i 增大时,折射角 γ 也相应增大,当入射角达到极大值($\pi/2$)时,所得到的折射角 γ_c 称为临界折射角。

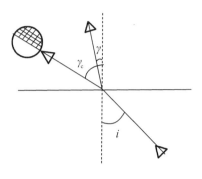

图 5-2-1 光的折射

显然从图中法线右边入射的光线从介质 B 进入介质 A 时,折射线都应落在临界折射角 γ_c 之内。在介质 A 中观察从介质 B 射入介质 A 的光线,小于临界角的区域为亮区,大于临界角的区域为暗区。临界折射角 γ_c 的大小和折射率有简单的函数关系,当固定介质 B 时,临界折射角 γ_c 的大小与介质 A 的特性有关。

测量折光率的阿贝折射仪就是根据临界角的原理设计的。测量物质的折射(光)率能定量地分析溶液的组成及检验物质的纯度。折射率与物质内部的电子运动状态有关,所以也用于结构化学方面的测定。

WAY-2S 数字阿贝折射仪通过角度-数字转换部件将角度量转换为数字量,通过数字显示被测样品的折射率或锤度(蔗糖溶液质量分数),易于读数。仪器如图 5-2-2 所示。WAY-2S 数字阿贝折射仪的使用方法如下:

(1)本仪器折射棱镜部件中有通恒温水结构,如需测定样品在某一特定温度下的折射率,仪器可外接恒温器,将温度调节到所需温度再进行测量。

(2)按下"POWER"电源开关,聚光照明部件中照明灯亮,同时显示窗显示 00000。有时显示窗先显示"—",数秒后显示 00000。

(3)打开折射棱镜部件,移去擦镜纸,仪器不使用时将这张擦镜纸放

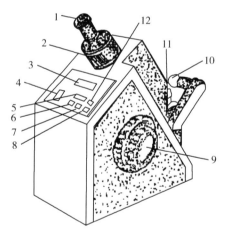

1—目镜;2—色散校正手轮;3—显示窗;4—POWER 电源;
5—READ 读数;6—BX-TC 经温度修正锤度;7—n_D 折射率;
8—BX 未经温度修正锤度;9—调节手轮;10—聚光照明部件;
11—折射棱镜部件;12—TEMP 温度。

图 5-2-2 WAY-2S 数字阿贝折射仪

在两棱镜之间,防止在关上棱镜时,可能留在棱镜上的细小硬粒弄坏棱镜工作表面。擦镜纸只需用单层。

(4)检查上、下棱镜表面,并用水或酒精小心清洁其表面。测定每一个样品后也要仔细清洁两块棱镜表面,因为留在棱镜上的少量的原来样品将影响下一个样品的测量准确度。

(5)将被测样品放在下面的折射棱镜的工作表面上。如图 5-2-3 所示,如样品为液体,可用干净滴管吸 1~2 滴液体样品放在棱镜工作表面上,然后将上面的进光棱镜盖上。如样品为固体,则固体样品必须有一个经过抛光加工的平整表面。测量前需将抛光表面擦净,并在下面的折射棱镜上表面上滴 1~2 滴折射率比固体样品折射率高的透明的液体(如溴代萘),然后将固体样品抛光面放在折射棱镜的工作表面上,使其接触良好。测固体样品时不需将上面的进光棱镜盖上。

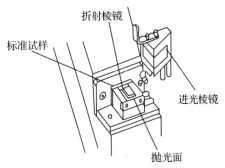

图 5-2-3 棱镜

(6)旋转聚光照明部件的转臂和聚光镜筒,使上面的进光棱镜的进光表面(测液体样品)或固体样品前面的进光表面(测固体样品)得到均匀照明。

(7)通过目镜观察视场,同时旋转调节手轮,使明暗分界线落在交叉线视场中。如从目镜中看到视场是暗的,可将调节手轮逆时针旋转;若看到视场是明亮的,则将调节手轮顺时针旋转。明亮区域在视场的顶部,在明亮视场情况下可旋转目镜,调节视度看清晰交叉线。

(8)旋转目镜方缺口里的色散校正手轮,同时调节聚光镜位置,使视场中明暗两部分具有良好的反差以及明暗分界线具有最小的色散。

(9)旋转调节手轮,使明暗分界线准确对准交叉线的交点,如图 5-2-4 所示。

(10)按"READ"读数显示键,显示窗中"00000"消失,显示"—"数秒后"—"消失,显示被测样品的折射率。可先选定测量方式,再按"READ"读数显示键,显示窗就按预先选定的测量方式显示。"BX"为未经温度修正的锤度,"BX-TC"为经温度修正的锤度,"nD"为折射率。

图 5-2-4 目镜视野

(11)检测样品温度,可按"TEMP"温度显示键,显示窗将显示样品温度。除了按"READ"键后,显示窗显示"—"时,按"TEMP"键无效,在其他情况下都可以对样品进行温度检测。显示为温度时,再按"nD""BX-TC"或"BX"键,显示的将是原来的折射率或锤度。为了区分显示值是温度还是锤度,在温度前加"t"符号,在"BX-TC"锤度前加"C"符号,在"BX"锤度前加"b"符号。

(12)样品测量结束后,必须用酒精或水(样品为糖溶液)小心清洁。仪器使用前后及更换样品时,必须先清洗、擦净折射棱镜系统的工作表面。

(13)仪器应避免强烈振动或撞击,防止光学零件震碎、松动而影响精度。本仪器严禁测试腐蚀性较强的样品。仪器聚光镜是用塑料制成的,为了防止带有腐蚀性的样品对它的表面产生破坏,必要时用透明塑料罩将聚光镜罩住。

仪器3　数字式差压计

压力是描述体系状态的重要参数之一。工程上把垂直、均匀作用在物体单位面积上的力称为压力,而物理学中则把垂直作用在物体单位面积上的力称为压强。在国际单位制中压强的单位为牛顿/米^2,取帕斯卡,符号为 Pa。物理概念就是 1 牛顿的力作用于 1 平方米的面积上所形成的压强(压力)。

由于地球上总是存在大气压力,当系统的压力低于大气压时,通常称为真空系统,可用真空度来表示该系统的压力。

绝对压力:实际存在的压力。

相对压力:和大气压力相比较得出的压力,又称表压力,一般压力表测出的是绝对压力和大气压力的差值。

正压力:绝对压力高于大气压力时的相对压力。

负压力:绝对压力低于大气压力时的相对压力,简称"负压",又名"真空"。差值的绝对值称为"真空度"。

差压力:任意两个压力相比较,其差值称为差压力,简称"压差"。

压力单位名称列于表 5 – 3 – 1 中。

表 5 – 3 – 1　压力单位名称符号

名称	符号	单位	换算关系
帕斯卡	Pa	牛顿/米^2	
大气压	atm		1 atm = 101 325 Pa
毫米汞柱(托)	Torr	mmHg	1 mmHg = 133.322 Pa
巴	Bar	10^6 达因/厘米^2	1 bar = 10^5 Pa
毫米水柱		mmH$_2$O	1 mmH$_2$O = 9.806 38 Pa

对于不同的压力范围和不同的精度要求,要选用不同的压力测量仪器。测量大气压有福廷式气压计、数字式气压计。测量压差有液柱差压计、数字式压差测量仪。测量真空的有数字式真空表。本文主要介绍两种用于测量压差的数字式差压计。

1. DPCY – 2C 型差压计

DPCY – 2C 型差压计又称 DPCY – 2C 型饱和蒸气压教学实验仪,如图 5 – 3 – 1 所示,用于测量待测系统与大气压的压力差值。它的特点是:采用全集成设计方案,使用集成电路芯片,选用精密差压传感器,将压力信号转换为电信号,微弱电信号经过低漂移、高精度的集成运算放大器放大后,再转换成数字信号;采用高亮度 LED 数字显示,数据直观,使用方便;仪器具有质量小、体积小、稳定性好、无汞污染、安全

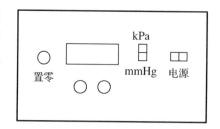

图 5 – 3 – 1　DPCY – 2C 型差压计面板

可靠等特点。

DPCY-2C 型差压计主要技术指标:电源电压 200~240 V,50 Hz;量程 -101~0 kPa;分辨率 0.01 kPa;环境温度 -20~40 ℃。

DPCY-2C 型差压计使用方法:

(1) 先使待测系统通大气,打开电源开关。

(2) 10 min 后按"置零"按钮,使压力显示框显示 0.00,通过面板上的双向开关"kPa/mmHg"选择数据单位。测量过程中不可轻易按"置零"按钮。

(3) 对系统抽真空,仪器数字显示框随时显示系统与大气压的压力差值。

(4) 注意保持仪器附近气流稳定,避免压力系统中的压力急剧变化。

DPCY-2C 型差压计校正:

(1) 当仪表使用一段时间后,若发现仪器不准时,用标准差压计对仪器进行校正。

(2) 将标准差压计与待校差压计接入同一压力系统,使压力系统连通大气,压力差应该为零,按下"置零"按钮,使待校差压计输出显示为零。

(3) 改变压力系统的压力至量程的 80% 附近,或较常使用的压力点。

(4) 按下"--"或"++"按钮,减小或增大显示值至标准差压计的显示值。

(5) 仪表的校正每年至少一次。

2. DMPY-2C 型微压差计

DMPY-2C 型微压差计又称为 DMPY-2C 型最大气泡法测定表面张力教学实验仪,如图 5-3-2 所示,用于测量待测系统与大气压的微小压差。仪器采用单片机测量系统,使用精密差压传感器,精度高,使用方便。

DMPY-2C 型微压差计主要技术指标:电源电压 200~240 V,50 Hz;量程为 -10~10 kPa;分辨率为 1 Pa;要求环境温度 -20~40 ℃。

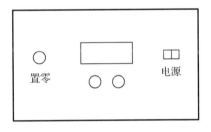

图 5-3-2　DMPY-2C 型微压差计

DMPY-2C 型微压差计使用方法:

(1) 将待测系统通大气,打开仪器电源开关,2 s 后正常显示。

(2) 预热 5 min 后按"置零"按钮,表示此时待测系统与大气压差为 0。

(3) 随着系统压力的变化,仪器跟踪显示待测系统的压力,如果待测系统的压力呈下降趋势,出现的极大值保留显示 1 s。压力差极小值与极大值出现的时间间隔不能太短,否则显示值将恒为极大值。

DMPY-2C 型微压差计校正:

(1) 当仪表使用一段时间后,或发现仪器不准时,用标准微压差计对仪器进行校正。

(2) 将标准微压差计与待校微压差计接入同一压力系统,使压力系统连通大气,压力差应该为零,按下"置零"按钮,使待校微压差计输出显示为零。

(3) 改变压力系统的压力至量程的 80% 附近,或较常使用的压力点。

(4) 按下"--"或"++"按钮,减小或增大显示值至标准微压差计的显示值。

(5) 仪表的校正每年至少一次。

仪器 4　电　导　率　仪

测量待测溶液电导的方法称为电导分析法。电导是电阻的倒数,因此电导值的测量实际上是通过电阻值的测量来换算的,也就是说电导的测量方法应该与电阻的测量方法相同。但在溶液电导的测定过程中,当电流通过电极时,由于离子在电极上会发生放电,产生极化引起误差,故测量电导时要使用频率足够高的交流电,以防止电解产物的产生。另外,所用的电极镀铂黑是为了减少超电位,提高测量结果的准确性。电解质溶液的电导测量除可用交流电桥法外,目前多数采用电导仪进行测量,它的特点是查测量范围广,快速直读及操作方便,如配接自动电子电势差计后,还可对电导的测量进行自动记录。电导仪的类型很多,基本原理大致相同,这里以 DDS – 11A 和 DDS – 307 电导率仪为例介绍电导率仪的原理及使用方法。

1. 测量原理

电导率仪的工作原理如图 5 – 4 – 1 所示。

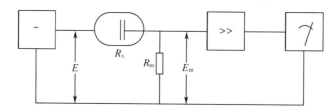

图 5 – 4 – 1　电导率仪的工作原理图

由图得出

$$E_m = \frac{ER_m}{R_m + R_x} = \frac{ER_m}{R_m + Q/\lambda}$$

式中,R_x 为液体电阻,即被玻璃固定在平行铂电极间的溶液的电阻;Q 为电导池常数(电极常数,$Q = X/\lambda$,X 为电极板间距,λ 为电极板面积);R_m 为分压电阻;E_m 为 R_m 两端的电压降;E 为音频标准电压。

由于 $E_m \ll R_x$,因此有电导率 γ 为

$$\gamma = \frac{QE_m}{ER_m}$$

由上式可见,当 E、R_m、Q 为常数时,γ 与 E_m 成正比,所以通过测量 E_m 的大小,可以测量出液体电导率的大小。

为降低极化作用造成的附加误差,测量信号 E 采用交流电,本机振荡产生的低频(约 140 Hz)及高频(约 1 100 Hz)两个频率分别作为低电导率测量及高电导率测量的信号源频率。振荡器用变压器耦合输出,因而信号 E 不随 R_x 变化。

放大后的信号经检波后,由刻有电导率读数的 0 ~ 1 mA 电流表指示出被测值,因为测

量信号是交流电,因而电极极片间及电极引线间均出现了不可忽视的分布电容 C_0(大约 60 pF),电导池则有电抗存在,因此,将电导池视做纯电阻来测量则存在比较大的误差,特别是 $0 \sim 10^{-7}\ \Omega^{-1} \cdot cm^{-1}$ 低电导率范围里,此项影响较显著,需采用电容补偿来消除,其原理见图 5-4-2。

信号源输出变压器的次级有两个输出信号 E_1 及 E_2,E_1 作为电容的补偿电源。E_1 与 E_2 的相位相反,所以由 E_1 引起的电流 I_1 流经 R_m 的方向与测量信号 I 流经 R_m 的方向相反。测量信号 I 中包括通过纯电阻 R_x 的电流和渡过分布电容 C_0 的电流。调节 K_6 可以使 I 与流过 C_0 的电流振幅相等,使它们在 R_m 上的影响大体抵消。

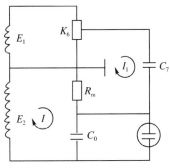

图 5-4-2　电容补偿原理图

2. DDS-11A 型数字电导率仪使用方法

DDS-11A 型数字电导率仪的面板如图 5-4-3 所示。

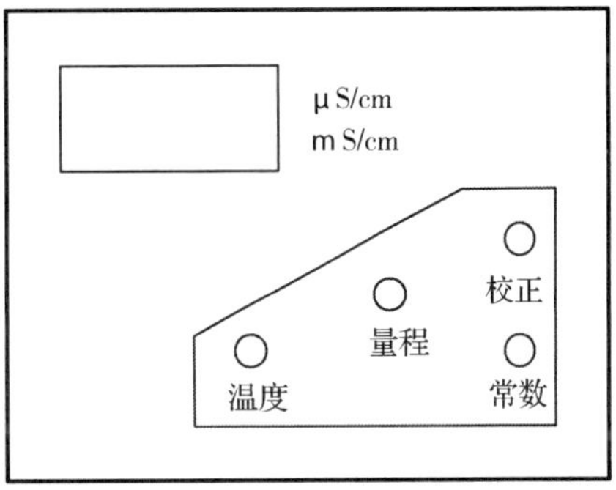

(1)将电源转换器的电源插头插入 220 V 电源插座,并将电源转换器的输出直流电源插头插入仪器的电源插座。

(2)将量程开关置于校正位置,温度旋钮置 25 ℃位置,电极常数旋钮置 1 位置,开机预热 10 min。调节校正调节器,使仪器读数为 199.9 $\mu S \cdot cm^{-1}$。

(3)用温度计测出被测介质温度后,把温度旋钮置于介质温度处。仪器的基准温度是 25 ℃。

(4)把常数旋钮置于与所使用电极的电池常数相一致的位置上。

图 5-4-3　DDS-11A 型数字电导率仪

①对 DJS-1 型铂电极,若常数为 0.95,则调节在 0.95 位置上;

②对 DJS-10 型铂电极,若常数为 11,则调节在 1.1 位置上。

(5)把量程开关置于校正位置,调节"校正"电位器使仪器显示 199.9 $\mu S \cdot cm^{-1}$ ±0.1。

(6)把量程开关置于所需的测量挡。如预先不知被测介质电导率的大小,应先把其置于最大电导率挡,然后逐挡选择适当范围,使仪器尽可能显示多位有效数字。此时仪器显示值即为溶液的电导率。如使用最大电导率挡量程时,仪器依然超量程,应换用常数为 10 的电导电极,数字显示值需乘以 10。

注意:在测量过程中如需重新校正仪器,只需将量程开关置于校正位置,即可重新校正仪器,而不必将电极插头拔出,也不必将电极从待测液中取出。仪器的校正挡与"温度"和"常数"调节器的位置有关,因此当温度位置和常数位置确定后,不能随意变动校正调节器的位置,否则影响测量精确度。

3. DDS-307 型电导率仪使用方法

(1)了解 DDS-307 型电导率仪的两极及调节器功能

DDS-307 型电导率仪仪器外形及各调节器功能如图 5-4-4 所示。

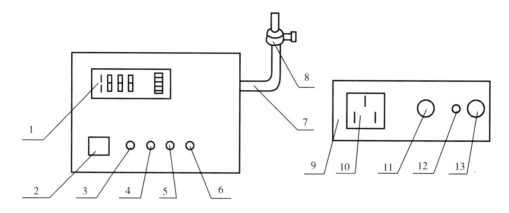

1—显示屏;2—电源开关;3—温度补偿调节器;4—常数选择开关;5—校正钮;6—量程开关;7—电极支架;
8—固定圈;9—后面板;10—三芯电源插座;11—保险丝管座;12—输入插口;13—电极插座。

图 5－4－4　DDS－307 型电导率仪仪器外形及各调节器功能

（2）电极的选用

按被测介质电导率（电阻率）的高低,选用不同常数的电导电极。

被测介质电导率小于 1 μS/cm（电阻率大于 1 MΩ·cm）,用常数为 0.01 的钛合金电极,测量时应加测量槽作流动测量。

测量介质电导率大于 100 μS/cm（电阻率 10 kΩ·cm）以上时,宜用常数为 1 或 10 的镀铂黑电导电极以增大吸附面,减少电极极化影响。

（3）调节"温度"旋钮

用温度计测出被测介质的温度后,把"温度"旋钮置于相应的介质温度刻度上。

注：若把旋钮置于 25℃ 线上,即为基准温度下补偿,即无补偿方式。

（4）"常数选择"开关的位置

若选用 0.01 cm^{-1} ±20%,常数的电极置于 0.01 处。

若选用 0.1 cm^{-1} ±20%,常数的电极置于 0.1 处。

若选用 1 cm^{-1} ±20%,常数的电极置于 1 处。

若选用 10 cm^{-1} ±20%,常数的电极置于 10 处。

（5）常数的设定及"校准"调节

量程开关置于"检查"挡：

①对 0.01 cm^{-1} 钛合金电极,电极选择开关置于 0.01 处;若常数为 0.009 5,则调节"校正"钮使显示值为 0.950。

②对 0.1 cm^{-1} 常数的 DJS－0.1C 型光亮电极,电极选择开关置于 0.1 处;若常数为 0.095 则调节"校正"钮使显示值为 9.50。

③对 1 cm^{-1} 常数的 DJS－1C 型电极,电极选择开关置于 1 处;若常数为 0.95 则调节"校正"钮使显示值为 95.0。

④对 10 cm^{-1} 常数的 DJS－10C 型电导电极,电极选择开关置于 10 处;若常数为 9.5 则调节"校正"钮使显示值为 950。

（6）把"量程"开关扳在测量挡,使显示值尽可能在 100～1 000。

(7)同时把电极插头插入插座,使插头的凹槽对准插座的凸槽,用食指按一下插头顶部,即可插入(拨出时捏住插头的下部,往上一拨即可),然后,把电极浸入介质,进行测量。

4. DDS-307A 型电导率仪及使用方法

DDS-307 型电导率仪外形及结构如图 5-4-5 所示,其液晶显示屏如图 5-4-6 所示。

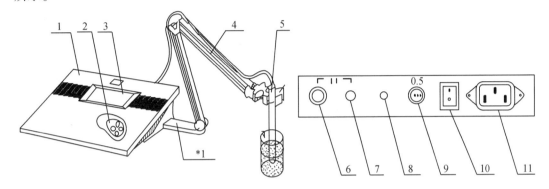

1—机箱(多功能电极架固定座已安装在机箱底部);2—键盘;
3—显示屏;4—多功能电极架;5—电导电极;6—测量电极插座;7.接地接口;
8—温度电极插座;9—保险丝(0.5 A);10—电源开关;11—电源插座。

图 5-4-5 DDS-307A 型电导率仪外形及结构

注:8.8.8.8 为电导率、TDS 测量数值;88.8 为温度显示数值。当仪器接上温度电极时,该温度显示数值为自动测量的温度值,即温度传感器反映的温度值。μS/cm、mS/cm 和 mg/L 为电导率、TDS 测量数值相应显示单位,℃为温度显示单位,℃闪烁时作为温度手动调节状态。测量、常数分别显示在相应工作状态。

图 5-4-6 DDS-307A 型电导率仪液晶显示屏

DDS-307A 型电导率仪使用方法:

(1)电源线插入仪器电源插座,仪器必须良好接地。

(2)按电源开关,接通电源,预热 30 min 后,进行测量。

(3)测量:

①应根据测量范围(参照表 5-4-1)选择相应常数的电导电极。

表 5-4-1　电极常数的选择

测量范围/($\mu S \cdot cm^{-1}$)	推荐使用的相应常数的电导电极
0~2	0.01、0.1
0~200	0.1、1.0
200~2 000	1.0
2 000~20 000	1.0、10
20 000~100 000	10

②电极常数的设置方法：

目前，电导电极的电极常数为0.01、0.1、1.0、10四种不同类型，每种类电极具体的电极常数值，制造厂均粘贴在每支电导电极上，可根据电极上所标的电极常数值调节仪器。按三次模式键，此时为常数设置状态，有"常数"二字显示，在温度显示数值的位置有数值闪烁显示，按"△"或"▽"键，闪烁数值显示在10、1、0.1、0.01程序转换。如果知道电导电极常数为1.025，则选择"1"并按"确认"键，此时在电导率、TDS测量数值的位置有数值闪烁显示，按"△"或"▽"键，闪烁数值显示在1.200~0.800范围内变化。如果知道电导电极常数为1.025，按"△"或"▽"键将闪烁数值显示为"1.025"，并按"确认"键，仪器回到电导率测量模式，至此校准完毕（电极常数为上下两组数值的乘积）。

③温度补偿的设置：

当仪器接上温度电极时，该温度显示数值为自动测量的温度值，即温度传感器反映的温度值，仪器根据自动测量的温度值进行自动温度补偿；当仪器不接温度电极时，该温度显示数值为手动设置的温度值，在温度值手动校准功能模式下（按"模式"键二次），可以按"△"或"▽"键手动调节温度数值上升、下降并按"确认"键，确认所选择的温度数值，使选择的温度数值为待测溶液的实际温度值，此时，测量得到的将是待测溶液经过温度补偿后折算为25 ℃下的电导率值。

如果"温度"补偿选择的温度数值为"25" ℃时，那么测量的将是待测溶液在该温度下未经补偿的原始电导率值。

④测量方法：

常数、温度补偿设置完毕，就可以直接进行测量，当测量过程中显示值为"1—"时，说明测量值超出量程范围，此时，应按"△"键，选择大一挡量程，最大量程为10 ms/cm或1 000 mg/L；当测量过程中，显示值为"0"时，说明测量值小于量程范围，此时，应按"▽"键，选择小一挡量程，最小量程为20 μS/cm或10 mg/L。

仪器5　旋　光　仪

许多物质具有旋光性，当平面偏振光线通过具有旋光性的物质时，它们可以将偏振光的振动面旋转某一角度。使偏振光的振动面向左旋的物质称为左旋物质，向右旋的称为右旋物质。若面向光源观察，向右偏转的角称为旋光角α，方向和大小与该物质分子立体结构有关,也受光的波长及物质温度的影响。通过测定旋光度的方向和大小,可以辅助鉴定物质,辅助判定有机物的分子立体构型,也用于测定物质的浓度。

根据我国国家标准 GB 3102.8—1993,比旋光度(Specific Optical Rotatory)的 SI 单位又定义为质量旋光本领 α_m(Mass Optical Rotatory Power)和摩尔旋光本领 α_n(Molar Optical Rotatory Power)。

质量旋光本领 α_m 定义为

$$\alpha_m = \alpha A/m \tag{5-5-1}$$

式中,m 为旋光性组元在截面积 A 的线性偏振光束途径中的质量;α 为旋光角,即平面偏振光通过旋光性介质,面向光源观察向右偏转的角(向左为 $-\alpha$);α_m 的 SI 单位是 $rad \cdot m^2 \cdot kg^{-1}$。

摩尔旋光本领 α_n 定义为

$$\alpha_n = \alpha A/n \tag{5-5-2}$$

式中,n 为旋光性组元在截面积 A 的线性偏振光束途径中的物质的量;α 为旋光角;α_n 的 SI 单位是 $rad \cdot m^2 \cdot mol^{-1}$。

由于旋光性物质溶液的旋光角与溶剂有关,故表示 α 值时,除注明温度、波长外还须注明溶剂。如果未注明,一般指水溶液。

测定物质旋光度的仪器称为旋光仪。目前在实验室通常使用两种类型的旋光仪:一种是实验者通过观察视野光线明暗,旋转一定角度,从刻度盘上读出旋光度;另一种是通过光电检测,数字显示旋光度。

1. WXG-4 型旋光仪

WXG-4 型旋光仪利用检偏镜来测量旋光角,光学系统如图 5-5-1 所示,它由起偏镜、检偏镜、两块辅助棱镜等组成。

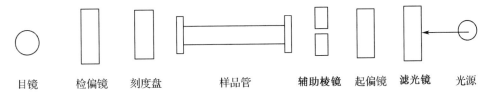

目镜　　检偏镜　刻度盘　　样品管　　辅助棱镜　起偏镜　滤光镜　光源

图 5-5-1　WXG-4 型旋光仪光学系统

光源为特制的钠光灯泡,为单色黄光(波长 589.3 nm)。由光源发出的光经滤光镜、起偏镜产生单一偏振光,通过检偏镜,如起偏镜、检偏镜的偏振面相互平行,光线可全部通过;如偏振面相互垂直,光线就完全不能通过;当两偏振面由相互垂直转向平行时(即由 90°→0°时),则通过光的强度由小变大。仪器中的起偏镜和辅助棱镜是固定的,而检偏镜与刻度盘连接并与刻度盘一起旋转调整角度,旋光角便可从刻度盘读出。

如果没有辅助棱镜,旋转检偏镜使其与起偏镜的偏振面相互垂直,则视场黑暗。在旋光管中盛旋光性溶液后,由于旋光性物质可使通过的偏振光的偏振面转动一个角度,则有部分光通过,视场稍明亮些,再旋转检偏镜适当角度,使视场又变为暗,所旋检偏镜的角度即是该旋光性物质的旋光角 α。面对光源观察,偏振方向顺时针方向旋转称右旋,反之,称左旋。

视场由亮变暗没有明显标志,不易准确测量。故在起偏镜后增加两个相互平行而分列两旁的辅助棱镜,并使辅助棱镜的偏振面与起偏镜的偏振面成适当角度 θ,如图 5-5-2 所示,则当旋光管中未盛旋光性物质时,有以下三种情况:

① 当检偏镜的偏振面与起偏镜的偏振面相互垂直时,视场内出现中间暗、两侧亮的情况,如图 5-5-2(a)所示;

② 当检偏镜的偏振面与辅助镜的偏振面相互垂直时,视场内出现中间亮、两侧暗,如图 5-5-2(b)所示。

③ 当检偏镜的偏振面与起偏镜、辅助棱镜两者的偏振面夹角 θ 的等分线相互垂直或平行时,则视场内三部分亮度(暗度)相等,如图 5-5-2(c)所示。

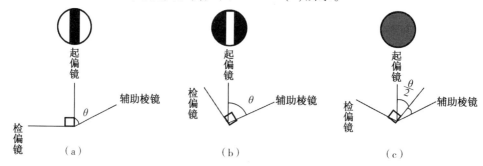

图 5-5-2 起偏镜、检偏镜与辅助棱镜之间的偏振面角度与视场图

如果旋光管中无旋光溶液时,转动检偏镜至视场亮度相等,记下其位置,然后在旋光管中盛旋光溶液,再旋转检偏镜至视场亮度相等,则所旋转的角度即是该旋光物质的旋光角 α。

旋光管有 10 cm 及 20 cm 两种,可视样品旋光能力选取合适的管长。旋光角的大小和管长成正比,与溶液中所含旋光性物质的浓度成正比。旋光角还与温度、入射光波长有关。

WXG-4 型旋光仪使用方法如下:

(1)图 5-5-3 所示的是 WXG-4 型旋光仪,将仪器接于 220 V 交流电源,开启电源开关,约 5 min 后钠光灯发光正常,即可开始测量;

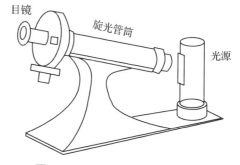

图 5-5-3 WXG-4 型旋光仪

(2)检查仪器零点,在旋光管中充满蒸馏水,调至视场内三分暗度相等,此时读出的值即为零点偏差值;

(3)选取长度合适的旋光管,注满待测溶液,若有小气泡(在较高温测量时应留一小气泡),应将气泡赶至旋光管的凸处;

(4)转动刻度盘,当视场亮度(暗度)一致时,从刻度盘上读数;

(5)旋光角与温度有关,测量时可以采用恒温措施。

2. WZZ-2B 自动旋光仪

WZZ-2B 自动旋光仪的原理和普通的光学旋光仪一样,其主要的使用方法如下:

(1)操作前的准备

①温度对旋光度影响不大的测试品,一般可在室温测定,如测定对温度有严格要求的测试品,在测定前应将仪器及测试样品置于规定温度的恒温箱内至少2 h,使温度恒定,否则会造成误差。测定温度为20±2 ℃。

②接通电源前,检查样品室内有无异物。

(2)操作方法

①打开仪器电源开关,钠光灯应启辉,但发光不稳,至少预热10 min,使之发光稳定。

②屏幕显示如下:欢迎使用WZZ-2B自动旋光仪。

③过几秒后,屏幕自动跳到设置界面,进入测量模式之前确定已转到直流状态,观察钠光灯亮度后再进入测量。

④按"⏎"键进入测量界面。

⑤仪器操作面板如图5-5-4所示。

图5-5-4　WZZ-2B自动旋光仪操作面板

操作面板左侧的显示屏可显示三组测量数据,下方为实测数值,等三组数据测量完毕后,此时显示的是三组数据的平均值。等显示数值不动后,按"清零"键进行清零。

⑥自动测量:按"自测"键,仪器就会自动测量三组(每组间电机正转0.5度左右)并在屏幕上显示平均值。若想重新测量,可直接按"自测"键。

⑦手动测量:按"手测"键,然后松开按键,仪器在测量一次后停下,等再次按键,可重复该动作,直至测量次数满三次,满三次后,若继续按"手测"键,屏幕会被清掉,在第一组位置显示被测数据。

⑧将装有注射用水或其他空白溶液的试剂的旋光管放入样品室,盖上箱盖按"清零"键,显示0读数。旋光管中若有气泡,应先让气泡浮在凸颈处,通光面两端的雾状水滴,应用软布擦干。旋光管螺帽不宜旋得过紧,以免产生应力,影响读数。旋光管安放时注意标记的位置和方向。

⑨取出旋光管,清洗干净后将待测样品注入旋光管,按相同的位置和方向放入样品室内,盖好箱盖,仪器将自动显示出该样品的旋光度。

⑩如样品超出测量范围,仪器在±45°处来回振荡,此时取出旋光管,仪器即自动转回零位。解决办法是稀释样品后重测。

⑪每次手动测量前,均需校零,如有误差按"清零"键。

⑫作好仪器使用记录及维护保养。

3. WZZ-1 自动旋光仪

将仪器电源插头插入 220 V 交流电源,然后打开电源开关,需经 5 min 钠光灯预热,使之发光稳定。

打开直流开关,若直流开关扳上后,钠光灯熄灭,则再将直流开关上下重复扳动 1～2 次,使钠光灯在直流下点亮为正常。

打开示数开关,将装有蒸馏水或其他空白溶剂的试管放入样品室,盖上箱盖,调节指针到零刻度。注意:试管中若有气泡,应先让气泡浮在凸颈处;通光面两端的雾状水滴用软布揩干。试管螺帽不宜旋得过紧,以免产生应力,影响读数。试管安放时应注意标记的位置和方向。

取出试管,将待测样品注入试管,按相同的位置和方向放入样品室内,盖好箱盖。示数盘将自动转出该样品的旋光度。示数盘上红色示值为左旋(-),黑色示值为右旋(+)。

逐次按下复测按钮,重复读几次数,取平均值作为样品的测定结果。

仪器使用完毕后,应依次关闭示数、直流、电源开关。

钠灯在直流供电系统出现故障不能使用时,仪器也可在钠灯交流供电的情况下测试,但仪器的性能可能略有降低。

仪器 6 EM-2A 型数字式电子电位差计

EM-2A 型数字式电子电位差计采用数字显示,利用对消法测量原理,内置了可代替标准电池、精度极高的参考电压集成块,作为比较电压。仪器线路设计采用全集成器件,待测电动势与参考电压经过高精度的仪表放大器比较输出,达至平衡时即可知待测电动势。

仪器要求电源电压 190～240 V,50 Hz;环境温度为 -20～40 ℃;量程为 0～1.5 V;分辨率为 0.01 mV。仪器面板如图 5-6-1 所示。

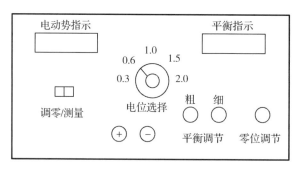

图 5-6-1 EM-2A 型数字式电子电位差计面板示意图

左上方为"电动势指示"6 位数码管显示窗口,右上方为"平衡指示"4 位数码管显示窗口。左边的开关可置于"调零"或"测量"挡,"调零"用于消除整个电路的零点误差,"测量"用于比较待测电动势和标准参考电压。右下角有 3 个多圈电位器,可进行"平衡调节",包括"粗""细"两个电位器和"零位调节"。"电位选择"为一个五挡的拨挡开关,可根据测量需要选挡。

两个标记为"+"和"-"的红黑接线柱接待测电池。

EM-2A 型数字式电子电位差计的使用方法及注意事项如下。

(1) 接通电源:打开电源开关,两组 LED 显示即亮,预热 5 min。

(2) 接线:待测电池按正负极性在红黑接线柱上接好。

(3) 选挡:"电位选择"分 0.3 V、0.6 V、1 V、1.5 V、2 V 挡,分别对应 0~0.3 V、0.3~0.6 V、0.6~1 V、1~1.5 V、1.5~2 V。两挡之间有一定的交叉,如待测电动势在 0.3 V 左右,选 0.3 V 和 0.6 V 挡均可以。选挡有两种方法:①根据估计的待测电动势,将"电位选择"拨至相应挡位;②任选一挡位,如"平衡指示"为"1999",则选此挡或降挡,如"平衡指示"为"-1999",则选此挡或升挡,再进一步调节电位器即可选出正确的挡位。

(4) 调零:将开关拨至"调零",调节"零位调节",使"平衡指示"数码显示稳定在零指示上,"电动势指示"此时显示为"------"。

(5) 测量:将开关拨至"测量",调节"平衡调节"的"粗""细"电位器,使"平衡指示"数码显示在零值附近。此时,等待"电动势指示"显示稳定下来,即为待测电动势值。注意,由于比较电路放大倍数很大和转换电脑精度很高,"电动势指示"和"平衡指示"数码显示在小范围内摆动属正常,摆动数值通常在 ±2 之间。

(6) 仪器不要放在有强电磁场干扰的区域内。

(7) 因仪器精度高,测量时应单独放置,不可将仪器叠放,也不要用手触摸仪器外壳。

(8) 仪器的精度较高,每次调节后,"电动势指示"的数码显示须经过一段时间才可稳定下来。

(9) 测量完毕后,应将待测电池接线及时取下。

仪器 7　JK99 全自动表面张力仪

表面张力是液体尤其是表面活性剂水溶液的一种基本性质。JK99 全自动表面张力仪是一种用物理方法代替化学方法简单易行地测试表面张力的仪器,用其可以迅速、准确地测出各种液体的表面张力值。JK99 全自动表面张力仪兼容白金板法和白金环,此处主要介绍白金板法(俗称吊片法)。

(1) 打开仪器开关,启动计算机,调出(全自动张力仪.EXE)应用程序,主界面如图 5-7-1 所示。

(2) 在选项菜单中点击"连接"选项,连接计算机与仪器。

(3) 参数设置。在选项菜单中点击"设置..."选项,设置测试模式、板法参数、环法参数、传感器校正及速度,如图 5-7-2 所示。

(4) 返回软件主界面,使用软件界面的"清零"按钮进行软件清零。

(5) 清洗白金板,步骤为:

①用镊子夹取白金板,并用流水冲洗,冲洗时应注意与水流保持一定的角度,原则为尽量让水流洗干净板的表面并且不能让水流使板变形;

②用酒精灯烧白金板,酒精灯一般与水平面呈 45°,直到白金板变微红为止,时间为 20~30 s;

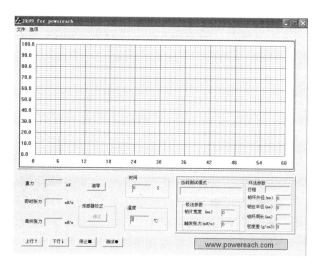

图 5-7-1　主界面示意图

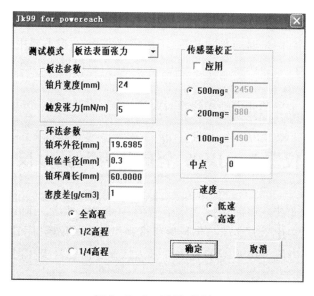

图 5-7-2　设置对话框

③注意事项:通常情况用水清洗即可,但遇有机液体或其他污染物用水无法清洗时,用丙酮清洗或用 20% HCL 加热 15 min 进行清洗,然后再用水冲洗,烧干即可。

(6) 在样品皿中加入测量液体,擦干样品皿外壁,在升降平台上垫上垫圈,将烧杯置于垫圈上。

(7) 准备就绪后,按"测试"开始记录,仪器会自动绘制整个表面张力值的变化曲线,数据记录完成后将整台曲线显示在屏幕上。可以记录表面张力值,也可以选择文件→"另存为…"存储实验结果。

(8) 重复性操作的方法为:按停止键,等表面张力仪样品台下降停止后,重新按测试键测试。

仪器 8　JS94H 微电泳仪

1. 启动

打开计算机和微电泳仪电源，进入 js94h 子目录，运行子目录中的 dh.exe，进入主界面后点击"Option"菜单中的"Connect"选项，出现"Connect Ok"，表明计算机与仪器的通信沟通成功。微电泳仪应用程序主界面如图 5-8-1 所示。

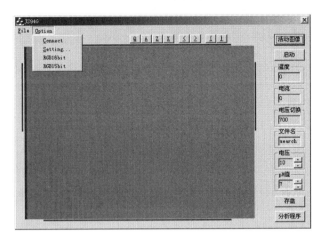

图 5-8-1　微电泳仪应用程序主界面图

2. 调焦与定位

用去离子水冲洗电泳杯和十字标，将被测样品注入电泳杯，插入十字标后洗涤数次，并让十字标充分湿润；取 0.5 mL 样品注入电泳杯，倾斜电泳杯，缓缓插入十字标，将含有十字标的表面接近镜头，细心观察，不要产生气泡。擦拭干电泳杯的外壁，将电泳杯平稳放入样品槽，轻轻按到底，切忌重压，如图 5-8-2 所示。

图 5-8-2　插入十字标

点击程序主界面上的"活动图像"按钮，调节电泳仪上的"左右调节螺杆""焦距调节螺杆"和"上下调节螺杆"旋钮，直到在计算机屏幕看到清晰的十字图像，如图 5-8-3 所示。

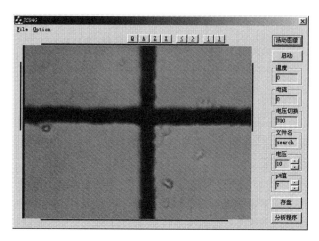

图 5-8-3　十字标调焦示意图

3. 采样测试

用去离子水冲洗电泳杯和电极,然后将被测样品注入电泳杯,插入电极后洗涤数次,并让电极装置充分湿润。接着取 0.5 mL 样品注入电泳杯,缓缓插入电极装置,细心观察,不要产生气泡。擦拭干电泳杯的外壁,将电泳杯平稳放入样品槽,连上电极连线,如图 5-8-4 所示。

图 5-8-4　电极连接示意图

点击"活动图像",调节所需电压,设置文件名,输入样品 pH 值,按"启动",图像上颗粒会随电极的切换左右移动,使用快捷键(Q:向上移动取景框;A:向下移动取景框;Z:向左移动取景框;X:向右移动取景框)调节,使待测颗粒处于取景框内,按"存盘",程序将截取图像供分析、计算时使用。

4. 分析

点击"分析程序"进入分析计算子程序界面,点击"开始",并输入文件名,然后系统将调出相应的图像和数据供用户分析,如图 5-8-5 所示。

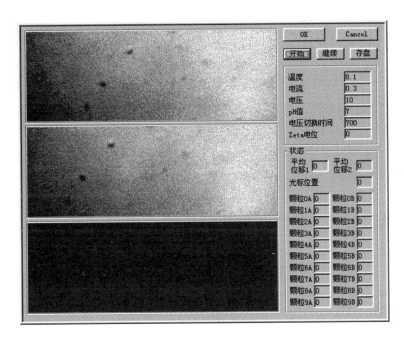

图 5-8-5　分析计算主程序界面图

在屏幕左侧有三个长方形的区域分别为定标分析区#1、#2、#3,右侧由上至下有三个区域,第一个是操作区,第二个是环境参数区,第三个是定标数据区。首先在分析区 #1 内确认一个颗粒,点击鼠标确认,在定标数据区内的颗粒 0A 位置将显示所确认的位置数据,然后根据颗粒位置的相关性,在分析区 #2 中确认同一颗粒的相同位置(即分析区 #1 内所确认的颗粒),其位置数据显示在定标数据区内的颗粒 0B 后,至此获得第一组数据;然后在分析区 #1 内再确认其他颗粒,用同样方法获得第二组数据,依此类推,可获得至多十组数据,然后按"存盘",输入颗粒电荷极性后,系统将给出颗粒的电泳淌度和 ξ 电势。如再按"继续"键,系统将调出第二组图像供用户分析。按"OK"退出分析计算子程序,回到主界面,进行下一项测量。实验结束以后,可以利用目录下的 dhprint.exe 程序打印实验数据和图像。

附录　常用数据表

附表-1　SI基本单位

物理量	单位名称	单位代号	
		国际	中文
长度	米(meter)	m	米
质量	千克(kilogram)	kg	千克
时间	秒(second)	s	秒
电流	安培(Ampare)	A	安
热力学温度	开尔文(Kelvin)	K	开
物质的量	摩尔(mole)	mol	摩
发光强度	坎德拉(candela)	cd	坎

附表-2　SI的一些导出单位

物理量	名称	代号		用国际制基本单位表示的关系式
频率	赫兹	Hz	赫	s^{-1}
力	牛顿	N	牛	$m \cdot kg \cdot s^{-2}$
压力	帕斯卡	Pa	帕	$m^{-1} \cdot kg \cdot s^{-2}$
能、功、热	焦耳	J	焦	$m^2 \cdot kg \cdot s^{-2}$
功率、辐射通量	瓦特	W	瓦	$m^2 \cdot kg \cdot m^{-3}$
电量、电荷	库伦	C	库	$s \cdot A$
电位、电压、电动势	伏特	V	伏	$m^2 \cdot kg \cdot s^{-3} \cdot A^{-1}$
电容	法拉	F	法	$m^{-2} \cdot kg^{-2} \cdot s^4 \cdot A^2$
电阻	欧姆	Ω	欧	$m^2 \cdot kg \cdot s^{-3} \cdot A^{-1}$
电导	西门子	S	西	$m^{-2} \cdot kg^{-1} \cdot s^3 \cdot A^2$
磁通量	韦伯	Wb	韦	$m^2 \cdot kg \cdot s^{-2} \cdot A^{-1}$
磁感应强度	特斯拉	T	特	$kg \cdot s^{-2} \cdot A^{-1}$
电感	亨利	H	亨	$m^2 \cdot kg \cdot s^{-2} \cdot A^{-2}$
光通量	流明	lm	流	$cd \cdot sr$
光照度	光照度	lx	勒	$m^{-2} \cdot cd \cdot sr$
黏度	帕斯卡秒	Pa·s	帕·秒	$m^{-1} \cdot kg \cdot s^{-1}$
表面张力	牛顿每米	N/m	牛/米	$kg \cdot s^{-2}$
热容量、熵	焦耳每开	J/K	焦/开	$K^2 \cdot kg \cdot s^{-2} \cdot K^{-1}$
比热	焦耳每千克每开	J/(kg·K)	焦/(千克·开)	$m^2 \cdot s^{-2} \cdot K^{-1}$
电场强度	伏特每米	V/m	伏/米	$m \cdot kg \cdot s^{-3} \cdot A^{-1}$

附录 常用数据表

附表-3 物理化学常数

常数名称	符号	数值	单位(SI)
真空光速	C	2.997 924 58	10^{-8} 米·秒$^{-1}$
基本电荷	e	1.602 189 2	10^{-19} 库仑
阿伏伽德罗常数	N_A	6.022 045	10^{23} 摩$^{-1}$
原子质量单位	M_u	1.660 565 5	10^{-27} 千克
电子静质量	m_e	9.109 534	10^{-31} 千克
质子静质量	m_p	1.672 648 5	10^{-27} 千克
法拉第常数	F	9.648 456	10^4 库仑·摩$^{-1}$
普朗克常量	h	6.626 176	10^{-34} 焦耳·秒
气体常数	R	8.314 41	焦耳·度$^{-1}$·摩$^{-1}$
波尔兹曼常数	k	1.380 662	10^{-23} 焦耳·度$^{-1}$
重力加速度	g	9.806 65	米·秒$^{-2}$

附表-4 水的表面张力 (单位:mN/m)

温度/℃	表面张力	温度/℃	表面张力	温度/℃	表面张力
15	73.49	21	72.59	27	71.66
16	73.34	22	72.44	28	71.50
17	73.19	23	72.28	29	71.35
18	73.05	24	72.13	30	71.18
19	72.90	25	71.97	31	70.38
20	72.75	26	71.82	32	69.56

附表-5 水的绝对黏度($\times 10^{-3}$) (单位:Pa·s)

温度/℃	0	1	2	3	4	5	6	7	8	9
0	1.787	1.728	1.671	1.618	1.567	1.519	1.472	1.428	1.386	1.346
10	1.307	1.271	1.235	1.202	1.169	1.139	1.109	1.081	1.053	1.027
20	1.002	0.977 9	0.954 3	0.932 5	0.911 1	0.890 4	0.870 5	0.851 3	0.832 7	0.814 8
30	0.797 5	0.780 8	0.764 7	0.749 1	0.734 0	0.719 4	0.705 2	0.691 5	0.678 3	0.665 4
40	0.652	0.640 8	0.629 1	0.617 8	0.606 7	0.596 0	0.585 6	0.575 5	0.565 6	0.556 1

附表-6　不同温度下液体的密度　　　　　　　　　　　　　　（单位：g/cm³）

温度/℃	水	乙醇	苯	汞	环己烷	乙酸乙酯	丁醇
5	0.999 9	0.808 2	—	13.583	—	0.918 6	0.820 4
6	0.999 9	0.801 2	—	13.581	0.790 6	—	—
7	0.999 9	0.800 3	—	13.578	—	—	—
8	0.999 8	0.799 5	—	13.576	—	—	—
9	0.999 8	0.798 7	—	13.573	—	—	—
10	0.999 7	0.797 8	0.887	13.571	—	0.912 7	—
11	0.999 6	0.797 0	—	13.568	—	—	—
12	0.999 5	0.796 2	—	13.566	0.785 0	—	—
13	0.999 4	0.795 3	—	13.563	—	—	—
14	0.999 2	0.794 5	—	13.561	—	—	—
15	0.999 1	0.793 6	0.883	13.559	—	—	—
16	0.998 9	0.792 8	0.882	13.556	—	—	—
17	0.998 8	0.791 9	0.882	13.554	—	—	—
18	0.998 6	0.791 1	0.881	13.551	0.783 6	—	—
19	0.998 4	0.790 2	0.881	13.549	—	—	—
20	0.998 2	0.789 4	0.879	13.546	—	0.900 8	—
21	0.998 0	0.788 6	0.879	13.544	—	—	—
22	0.997 8	0.787 7	0.878	13.541	—	—	0.807 2
23	0.997 5	0.786 9	0.877	13.539	0.773 6	—	—
24	0.997 3	0.786 0	0.876	13.536	—	—	—
25	0.997 0	0.785 2	0.875	13.534	—	—	—
26	0.996 8	0.784 3	—	13.532	—	—	—
27	0.996 5	0.783 5	—	13.529	—	—	—
28	0.996 2	0.782 6	—	13.527	—	—	—
29	0.995 9	0.781 8	—	13.524	—	—	—
30	0.995 6	0.780 9	0.869	13.522	0.767 8	0.888 8	0.800 7

附表-7　水的折射率（钠光）

温度/℃	折射率	温度/℃	折射率	温度/℃	折射率
0	1.333 95	19	1.333 08	26	1.332 43
5	1.333 88	20	1.333 00	27	1.332 31
10	1.333 68	21	1.332 92	28	1.332 19
15	1.333 37	22	1.332 83	29	1.332 06
16	1.333 30	23	1.332 74	30	1.331 92
17	1.333 23	24	1.332 64		
18	1.333 16	25	1.332 54		

附表-8 一些液体的折射率(25 ℃)

名称	折射率	名称	折射率	名称	折射率
甲醇	1.326	乙酸乙酯	1.370	甲苯	1.494
水	1.332 50	正己烷	1.372	苯	1.498
乙醚	1.352	丁醇-1	1.397	苯乙烯	1.545
丙酮	1.357	氯仿	1.444	溴苯	1.557
乙醇	1.359	四氯化碳	1.459	苯胺	1.583
醋酸	1.370	乙苯	1.493	溴仿	1.587

附表-9 一些液体的蒸气压

物质的蒸气压 $p(\text{Pa})$ 按下式计算：

$$\lg p = A - \frac{B}{C+T} + D$$

式中，A、B、C 为常数；T 为温度，℃；D 为压力单位的换算因子，其值为 2.124 9。

名称	分子式	适用温度范围/℃	A	B	C
四氯化碳	CCl_4		6.879 26	1 212.021	226.41
氯仿	$CHCl_3$	-30~150	6.903 28	1 163.03	227.4
甲醇	CH_4O	-14~65	7.897 50	1 474.08	229.13
1,2-二氯乙烷	$C_2H_4Cl_2$	-31~99	7.025 3	1 271.3	222.9
醋酸	$C_2H_4O_2$	0~36	7.803 07	1 651.2	225
		36~170	7.188 07	1 416.7	211
乙醇	C_2H_6O	-2~100	8.321 09	1 718.10	237.52
丙醇	C_3H_6O	-30~150	7.024 47	1 161.0	224
异丙醇	C_3H_6O	0~101	8.117 78	1 580.92	219.61
乙醇乙酯	$C_4H_9O_2$	-20~150	7.098 08	1 238.71	217.0
正丁醇	$C_4H_{10}O$	15~131	7.476 80	1 362.39	178.77
苯	C_6H_6	-20~150	6.906 61	1 211.033	220.790
环己烷	C_6H_{12}	20~81	6.841 30	1 201.53	222.65
甲苯	C_7H_8	-20~150	6.954 64	1 344.80	219.482
乙苯	C_6H_{10}	-20~150	6.957 19	1 424.251	213.206

附表-10 标准电极电势及其温度系数(298.15 K)

电极	电极反应	$\varphi^{\ominus}/V$	$(d\varphi^{\ominus}/dT)/(mV \cdot K^{-1})$
Li^+/Li	$Li^+ + e^- = Li$	-3.045	-0.534
Na^+/Na	$Na^+ + e^- = Na$	-2.714	-0.772
Al^{3+}/Al	$Al^{3+} + 3e^- = Al$	-1.662	0.504
Mn^{2+}/Mn	$Mn^{2+} + 2e^- = Mn$	-1.180	-0.08
$OH^-/H_2,Pt$	$2H_2O + 2e^- = H_2 + 2OH^-$	-0.828 1	-0.834 2
Zn^{2+}/Zn	$Zn^{2+} + 2e^- = Zn$	-0.762 8	0.091
Gr^{3+}/Gr	$Gr^{3+} + 3e^- = Gr$	-0.744	0.468
S^{2-}/S	$S + 2e^- = S^{2-}$	-0.51	
Fe^{2+}/Fe	$Fe^{2+} + 2e^- = Fe$	-0.440	0.052
Ni^{2+}/Ni	$Ni^{2+} + 2e^- = Ni$	-0.250	0.06
$I^-/AgI,Ag$	$AgI + e^- = Ag + I^-$	-0.152	-0.248
Sn^{2+}/Sn	$Sn^{2+} + 2e^- = Sn$	-0.136	-0.282
Pb^{2+}/Pb	$Pb^{2+} + 2e^- = Pb$	-0.126	-0.451
$H^+/H_2,Pt$	$2H^+ + 2e^- = H_2(g)$	0.000	0.000
$Sn^{4+},Sn^{2+}/Pt$	$Sn^{4+} + 2e^- = Sn^{2+}$	0.15	
$Cu^{2+},Cu^{1+}/Pt$	$Cu^{2+} + e^- = Cu^+$	0.153	0.073
$Cl^-/AgCl,Ag$	$AgCl + e^- = Ag + Cl^-$	0.222 4	-0.658
Cu^{2+}/Cu	$Cu^{2+} + 2e^- = Cu$	0.337	0.008
$OH^{-1}/O_2,Pt$	$O_2(g) + 2H_2O + 4e^- = 4OH^-$	0.401	-0.44
Cu^+/Cu	$Cu^+ + e^- = Cu$	0.521	-0.058
$I^-/I_2,Pt$	$I_2 + 2e^- = 2I^-$	0.535 5	-0.148
$Fe^{3+},Fe^{2+}/Pt$	$Fe^{3+} + e^- = Fe^{2+}$	0.771	1.188
Ag^+/Ag	$Ag^+ + e^- = Ag$	0.799 1	-1.000
$H^+/O_2,Pt$	$O_2(g) + 4H^+ + 4e^- = 2H_2O$	1.229	-0.846
$Cr^{3+},Cr_2O_7^{2-},H^+/Pt$	$Cr_2O_7^{2-} + 14H^+ + 6e^- = 2Cr^{3+} + 7H_2O$	1.33	-1.263
$Cl^-/Cl_2,Pt$	$Cl_2 + 2e^- = 2Cl^-$	1.359 5	-1.26
Au^{3+}/Au	$Au^{3+} + 3e^- = Au$	1.498	
$Ce^{4+},Ce^{3+}/Pt$	$Ce^{4+} + e^- = Ce^{3+}$	1.61	
$SO_4^{2-},H^+/PbSO_4 \cdot PbO_2$	$PbO_2 + SO_4^{2-} + 4H^+ + 2e^- = PbSO_4 + 2H_2O$	1.682	-0.326
Au^+/Au	$Au^+ + e^- = Au$	1.691	
$F^-/F_2,Pt$	$F_2 + 2e^- = 2F^-$	2.87	-1.830

附表-11 无限稀释水溶液中离子摩尔电导率(298 K)

离子	$\Lambda_m^\infty \times 10^4$ /(S·m²·mol⁻¹)	离子	$\Lambda_m^\infty \times 10^4$ /(S·m²·mol⁻¹)	离子	$\Lambda_m^\infty \times 10^4$ /(S·m²·mol⁻¹)
H^+	349.65	$1/2\ Ca^{2+}$	59.47	$1/2\ SO_4^{2-}$	80.0
K^+	73.48	$1/3\ La^{3+}$	69.7	$1/2\ C_2O_4^{2-}$	74.11
Na^+	50.08	OH^{-1}	198	$1/3\ C_6H_5O_7^{3-}$	70.2
NH_4^+	73.5	Cl^{-1}	76.31	$1/4\ Fe(CN)^{4-}$	110.4
Ag^+	61.9	No_3^{-1}	71.42		
$1/2\ Ba^{2+}$	63.6	$C_2H_2O_2^{2-}$	40.9		

附表-12 不同温度下 KCl 水溶液的电导率 κ (单位:S·cm⁻¹)

$T/℃$	κ		
	0.01 mol·L⁻¹	0.02 mol·L⁻¹	0.10 mol·L⁻¹
15	0.001 147	0.002 043	0.010 48
16	0.001 173	0.002 294	0.010 72
17	0.001 199	0.002 345	0.010 95
18	0.001 225	0.002 449	0.011 43
19	0.001 251	0.002 449	0.011 43
20	0.001 278	0.002 501	0.011 91
21	0.001 305	0.002 553	0.011 91
22	0.001 332	0.002 606	0.012 15
23	0.001 359	0.002 659	0.012 39
24	0.001 386	0.002 712	0.012 64
25	0.001 431	0.002 765	0.012 88
26	0.001 441	0.002 873	0.013 37
27	0.001 468	0.002 873	0.013 37
28	0.001 496	0.002 927	0.013 62
29	0.001 524	0.002 981	0.013 87
30	0.001 552	0.003 036	0.014 12

附表-13 一些强电解质的活度系数

物质	浓度/mol·kg⁻¹			
	0.01	0.1	0.5	1
$AgNO_3$	0.90	0.734	0.536	0.429
$CuSO_4$	0.40	0.150	0.062	0.042 3
HCl		0.976	0.757	0.809
H_2SO_4		0.265 5	0.155 7	0.131 6
KBr		0.772	0.657	0.617
KCl		0.770	0.649	0.604
KNO_3		0.739	0.545	0.443
NH_4Cl		0.770	0.649	0.603
NaCl	0.903 2	0.778	0.681	0.657
NaOH		0.766	0.690	0.678
$ZnSO_4$	0.387	0.150	0.063 0	0.043 5

附表-14 常用参比电极的电势及温度系数

名称	体系	E/V^*	$(dE/dT)/(mV \cdot K^{-1})$
氢电极	Pt, H_2 ∣ H^+($a_{H^+}=1$)	0.000 0	
饱和甘汞电极	Hg, Hg_2Cl_2 ∣ 饱和 KCl	0.241 5	-0.761
标准甘汞电极	Hg, Hg_2Cl_2 ∣ 1 mol·L^{-1} KCl	0.280 0	-0.275
0.1 mol·L^{-1}甘汞电极	Hg, Hg_2Cl_2 ∣ 0.1 mol·L^{-1} KCl	0.333 7	-0.875
银-氯化银电极	Ag, AgCl ∣ 0.1 mol·L^{-1} KCl	0.290	0.3
氧化汞电极	Hg, HgO ∣ 0.1 mol·L^{-1} KOH	0.165	
硫酸亚汞电极	Hg, Hg_2SO_4 ∣ 1 mol·L^{-1} Hg_2SO_4	0.675 8	
硫酸铜电极	Cu ∣ 饱和 $CuSO_4$	0.316	0.7

注：* 25 ℃相对于标准氢电极(NHE)。

附表-15 不同温度下饱和甘汞电极(SCE)的电极电势

$T/℃$	φ/V^*	$T/℃$	φ/V^*
0	0.256 8	40	0.230 7
10	0.250 7	50	0.223 3
20	0.244 4	60	0.215 4
25	0.241 2	70	0.207 1
30	0.237 8		

注：* 25 ℃相对于标准氢电极(NHE)。

附表-16 甘汞电极的电极电势与温度的关系

甘汞电极*	φ/V
SCE	$0.241\ 2 - 6.61 \times 10^{-4}(t-25) - 1.75 \times 10^{-6}(t-25)^2 - 9 \times 10^{-10}(t-25)^3$
NCE	$0.280\ 1 - 2.75 \times 10^{-4}(t-25) - 2.50 \times 10^{-6}(t-25)^2 - 4 \times 10^{-9}(t-25)^3$
0.1NCE	$0.333\ 7 - 8.75 \times 10^{-5}(t-25) - 3 \times 10^{-6}(t-25)^2$

注：*SCE 为饱和甘汞电极；NCE 为标准甘汞电极；0.1NCE 为 0.1 mol·L^{-1}甘汞电极；相对于标准氢电极。

附表-17 有机化合物的密度与温度的关系

表中所列有机物的密度可用下列方程计算：

$$\rho_t = \rho_0 + 10^{-3}\alpha t + 10^{-6}\beta t^2 + 10^{-9}\gamma t^3$$

式中，ρ_0 为 0 ℃时的密度，g·cm^{-3}；ρ_t 为温度为 t 时的密度，g·cm^{-3}。

化合物	P_v	α	β	γ	温度范围/℃
四氯化碳	1.632 55	-1.941 0	-0.690		0~40
氯仿	1.526 43	-1.856 3	-0.560	-8.84	-53~+55
乙醚	0.736 29	-1.113 8	-1.237		0~70
乙醇*	0.785 06	-0.859 1	-0.56	-5	10~40
醋酸	1.072 4	-1.122 9	-0.005	-2.0	9~100
丙酮	0.812 48	-1.100	-0.858		0~50
乙酸乙酯	0.924 54	-1.168	-1.95	+20	0~40
环己烷	0.797 07	-0.887 9	-0.972	1.55	0~60

注：*0.785 06 为 25 ℃时的密度，利用上述方程计算时，温度项应该用 $T-25$。

参 考 文 献

[1] 傅献彩,沈文霞,姚天扬,等.物理化学(上册)[M].5版.北京:高等教育出版社,2005.
[2] 傅献彩,沈文霞,姚天扬,等.物理化学(下册)[M].5版.北京:高等教育出版社,2006.
[3] 天津大学物理化学教研室.物理化学(上、下册)[M].4版.北京:高等教育出版社,2001.
[4] 孙尔康,徐维清,邱金恒.物理化学实验[M].南京:南京大学出版社,1998.
[5] 北京大学化学系物理化学教研室.物理化学实验[M].3版.北京:北京大学出版社,1995.
[6] 复旦大学等.物理化学实验[M].3版.北京:高等教育出版社,2004.
[7] 郑传明,吕桂琴.物理化学实验[M].北京:北京理工大学出版社,2005.
[8] 刘廷岳,王岩.物理化学实验[M].北京:中国纺织出版社,2006.
[9] 杨冬花,武正簧.物理化学实验[M].徐州:中国矿业大学出版社,2005.
[10] 王玉峰,孙墨珑,张秀成.物理化学实验[M].哈尔滨:东北林业大学出版社,2006.
[11] 沈阳化工学院物理化学教研室.物理化学实验[M].大连:大连理工大学出版社,2006.
[12] 王兵,陶庆泰.天津工业大学物理化学讲义[M].天津:天津工业大学出版社,2006.
[13] 张美珍.聚合物研究方法[M].北京:中国轻工业出版社,2006.
[14] 王光信,张积树.有机电合成导论[M].北京:化学工业出版社,1997.
[15] 马淳安.有机电化学合成导论[M].北京:科学出版社,2002.
[16] 唐林,孟阿兰,刘红天.物理化学实验[M].北京:化学工业出版社,2008.
[17] 赵鹏,王维德,倪海霞.有机电化学合成[J].化工装备技术,2005,26(3):32-35.
[18] 刘志明,李琦,孙清瑞,等.龙葵红色素热降解化学动力学研究[J].食品工业科技,2009,30(6):309-311.
[19] 刘小玲,宁恩创,林莹,等.红龙果甜菜苷色素降解动力学研究[J].食品研究与开发,2008,29(2):44-47.
[20] 刘志明,吴也平,金丽梅.应用物理化学实验[M].北京:化学工业出版社,2009.
[21] 川合慧.铝阳极氧化膜电解着色及其功能膜的应用[M].北京:冶金工业出版社,2005.
[22] 暨调和,曾凌三,张国芝.建筑铝型材的阳极氧化和电解着色[M].长沙:湖南科学技术出版社,1994.
[23] 李云凯.金属材料学[M].北京:北京理工大学出版社,2006.
[24] 杨辉,卢文庆.应用电化学[M].北京:科学出版社,2001.
[25] 安茂忠,李丽波,杨培霞.电镀技术与应用[M].北京:机械工业出版社,2007.
[26] 冯辉,张勇,张林森.电镀理论与工艺[M].北京:化学工业出版社,2008.
[27] ONUCHUKWU A I, MSHELIA P B. The production of oxygen gas: A student catalysis experiment[J]. Journal of Chemical Education,1985,62(9):809.
[28] GOLDSTEIN J R, TSEUNG A C C. Kinetics of oxygen reduction on graphite/cobalt-iron

oxide electrodes with coupled heterogeneous chemical decomposition of H_2O_2 [J]. The Journal of Physical Chemistry, 1972, 76(24):3646-3656.

[29] 李德忠. 皂化反应动力学实验中几个问题的讨论[J]. 化学通报, 1992(9):53-55.

[30] 金家骏. 乙酸乙酯碱性水解的动力学盐效应[J]. 化学通报, 1974(4):28-31.

[31] 朱埗瑶, 赵国玺. 液体表(界)面张力的测定——滴体积法介绍[J]. 化学通报, 1981(6):21-26.